THE CHEMISTRY OF
TOBACCO AND TOBACCO SMOKE

L Symposium on ...
"

THE CHEMISTRY OF
TOBACCO AND TOBACCO SMOKE

Proceedings of the Symposium on the Chemical Composition of Tobacco
and Tobacco Smoke held during the 162nd National Meeting of the American
Chemical Society in Washington, D.C., September 12-17, 1971

Edited by Irwin Schmeltz

Eastern Marketing and Nutrition Research Division
U. S. Department of Agriculture
Philadelphia, Pennsylvania

Ⴔ PLENUM PRESS • NEW YORK-LONDON • 1972

CHEMISTRY

Library of Congress Catalog Card Number 72-76934
ISBN 0-306-30597-6

© 1972 Plenum Press, New York
A Division of Plenum Publishing Corporation
227 West 17th Street, New York, N. Y. 10011

United Kingdom edition published by Plenum Press, London
A Division of Plenum Publishing Company, Ltd.
Davis House (4th Floor), 8 Scrubs Lane, Harlesden, London,
NW10 6SE, England

Printed in the United States of America

PREFACE

The present volume comprises a compilation of papers presented as a Symposium on the Chemical Composition of Tobacco and
Tobacco Smoke during a meeting of the American Chemical Society
in Washington, D. C., September 12-17, 1971.

The Symposium was organized so as to cover, in the time
allotted, those aspects of tobacco research that are both pertinent and relevant to the most demanding problem facing researchers in the field today--that is the development of a less hazardous cigarette. The path to such an objective, however, is still
rather long and not easily traversed.

For example, in identifying the hazard associated with smoking, one must first know something of the chemical composition of
tobacco smoke, and moreover, how the smoke components arise from
the various leaf components. In addition, bioassays of smoke
fractions and components therein are necessary to identify noxious
substances, and to correlate biological activity with chemical
composition. Finally, to achieve the stated objective, methods
need to be developed for removing the identified hazards from
the smoke--whether they be by specially cultivating tobacco
plants, or by modifying tobacco smoke through the use of filters,
additives or similar devices.

The intent of the Symposium was to explore all the above
areas. Thus papers were presented which discussed various aspects
of the composition of tobacco leaf and how that composition might
be altered in the future. In papers on recent findings in the
chemical composition of tobacco smoke, methods for fractionating
the smoke, and resolving complex mixtures of smoke components by
the use of combined capillary column gas chromatography--mass
spectrometry were described. Pyrolysis studies which related
leaf and smoke components and which attempted to shed light on
the mode of origin of smoke components, including those considered
to be biologically active, were also taken up by the Symposium.
Discussions of methods for bioassaying smoke condensate, and
results obtained from utilizing such methods for a number of smoke
fractions and identifying the active chemical constituents in
these fractions comprised a significant portion of the Symposium.
Finally, with all the above as a foundation, the Symposium took

up the question of modifying the smoke and considered in detail the use of various filter systems and additives.

It is hoped, of course, that the Symposium papers, presented in this Volume, will make a contribution, by way of providing pertinent information to research workers in the field, to the objective stated earlier, of current tobacco research.

The editor wishes to thank associates and colleagues who helped, at various times, with the organization and presentation of the Symposium: Dr. R. L. Stedman and Dr. C. F. Woodward who suggested several of the speakers, Dr. J. B. Fishel and Dr. O. T. Chortyk who presided at sessions and all the speakers and participants who contributed their time and knowledge in the field. In addition, the cooperation of Dr. Emily L. Wick and the other officers of the Agricultural and Food Chemistry Division of the American Chemical Society must be acknowledged.

Irwin Schmeltz

CONTENTS

THE CHEMISTRY OF
TOBACCO AND TOBACCO SMOKE

RECENT TRENDS IN TOBACCO AND TOBACCO SMOKE RESEARCH

Helmut Wakeham

Philip Morris Research Center

Richmond, Virginia

The excuse for a research director to review before his scientific peers recent progress in the field of his own program does not come often. It is an opportunity not to be missed. In many ways it is more challenging than presenting the program to management who foots the bill. It has the advantage of putting one's own program into better perspective. Any research group tends to become so engrossed in its own work that it overlooks peripheral areas of work -- we see the trees and forget to look at the forest. Frankly, I am not sure my associates have appreciated my involvement in this symposium. In the course of my preparation I have suggested some gaps in our own program which they should be exploring.

In this review I shall refrain from ticking off new compounds recently found in tobacco or smoke, or listing new techniques in analysis and pyrolysis. I shall also not review all the chemical studies being carried out in connection with agricultural and curing investigations on tobacco. I propose rather to view broadly the field of tobacco and smoke chemistry as it pertains to the real life smoking situation. What chemistry is taking place as the smoker puffs away on his smoking product? In laboratories we often do simplified experiments under arbitrary and artificial conditions because it is easier to do things that way. We forget that these conditions may be quite unlike those experienced by human smokers, each of whom is enjoying his pipe or cigaret in different ways from his fellow smokers.

Chemists have been intrigued by the complexity of tobacco and tobacco smoke from the turn of this century. Their interest arose not only from the widespread use of tobacco, but also from the fact that, except for chewing tobacco, the product of interest is the smoke derived from the combustion of tobacco. Here we have a natural product containing hundreds of chemical constituents exposed to temperatures ranging up to over a thousand degrees Centigrade in the presence of varying amounts of oxygen. Where else, except possibly in complex living organisms, could the chemist find such an inexhaustible supply of chemical problems to keep him forever challenged? Small wonder that the study of tobacco and smoke chemistry has mushroomed even faster than the over-all growth of science.

The first challenge, of course, was analytical. What is in tobacco and smoke to give rise to the pleasurable experience of its use -- the taste, the aroma, the physiological and psychological sensations? Already by 1930 extensive analyses had been made of all categories of substances: alkaloids, carbohydrates, proteins, acids, hydrocarbons, sterols, phenols and polyphenols; (5) (40) and all of this without the sensitive instrumental methods we have today. In fact, it is interesting to note that in their 1959 Chemical Reviews of the "Constituents of Tobacco and Tobacco Smoke" Johnstone and Plimmer (21) were reluctant to accept "isolations and identifications" based on "R_f values, color tests, and ultra-violet spectra" instead of the "criteria of classical organic chemistry." In contrast with this restricted view we now have the sophisticated combined chromatographic, mass spectographic methods. Grob and Vollmin (13), for example, reported after one pass identification of 133 compounds in smoke.

Analysis of tobacco and smoke was, of course, greatly stimulated by the smoking and health controversy. A great search has been made for toxic substances in smoke in attempts to explain the statistical associations with disease. Some analysts with sophisticated methods are now finding chemicals in fractional nanogram quantities, often ignoring the fact that toxicity is a function of concentration. They are looking for the needle in the haystack, as it were, to prevent the cow from eating it.

Consequently tobacco chemists now have compiled a list of some 1350 identified chemical components in tobacco and smoke. Gas chromatographic scans indicate there are many more, probably over ten thousand, possibly even a hundred thousand. Considering the chemical nature of tobacco and the cigaret burning process, we should not be startled by such a prediction. We should rather recognize that simply finding more things in smoke may not be as useful in our quest for applicable knowledge as understanding the nature of the chemical processes involved in growing, curing, processing tobacco and producing pleasurable smoke.

Chemists in the tobacco industry are naturally interested in the smoke generating process. Because tobacco is a product of nature, and because of the tobacco production and marketing system, they have relatively little control of their raw material. Any hope they might have of controlling smoke composition for flavor or other reasons will depend largely on manipulating the cigaret, cigar, or pipe. Apart from filtration, which is another subject, the chemists are more or less limited to the process of converting tobacco into smoke.

It is instructive to take a broad look at the compositions of tobacco and of smoke. Of our 1350 chemical constituents, 440 are reported for tobacco alone, 510 are found in smoke, and about 400 are found in both. (42) It should be noted that many of those substances found only in tobacco or in smoke may simply not have been sought in the other by the investigator. Some we might expect to find in both, some in only one or the other state.

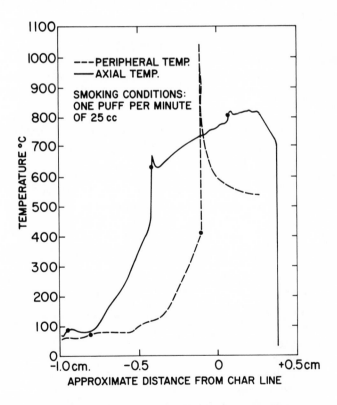

Figure 1. Temperature Profile at Axial and Peripheral Positions Behind the Tip of the Burning Cone. (2) (28)

Let us consider briefly the transfer of chemicals from tobacco to smoke. Those found in both states obviously are being driven from the tobacco into the smoke by the heat of the burning coal. The process involves evaporation and condensation.

Figure 1 charts the increase in temperature of tobacco behind the burning coal in a cigaret as a function of distance.[2] [28] It shows how both during puffs and between puffs the tobacco is getting hotter and hotter as the burning zone is approaching. Visual and thermogravimetric observations indicate that even as low as 300°C vapors from the tobacco are condensing to form smoke, at 450°C charring takes place, and around 600°C the tobacco is kindled and starts to burn.[31] Behind the coal there is a temperature gradient which is very sharp between puffs and less so during the puff when the stream of hot gases from the burning coal passes back through the tobacco rod. Substances which are highly volatile thus are readily distilled from the tobacco behind the hot coal. Less volatile substances being transferred must exhibit adequate vapor pressures and sufficient chemical stability so that the process will take place without chemical modification before evaporation. Let us consider some examples (Table 1).

In this case we would expect about one-third transfer to take place because about that fraction of the tobacco is burning during the puff. We find good transfers for low molecular weight volatile components, less for the higher boiling constituents, least for the large molecules.

Dotriacontane, despite its high boiling point and molecular size, transfers well because it is a stable hydrocarbon.

The presence of larger, higher molecular weight substances in smoke is not clear. Solanesol, which is a C-45 isoprenoid, sterols, sugars, leaf pigments, and even cellulose have all been reported in

Table 1.   Tobacco Chemicals Transferred to Mainstream Smoke.

COMPOUND	BP	% TRANSFER (APPROX.)*
Menthol	212° C	39
Nicotine	245° C	24
Glycerol	d. 290° C	22
Neophytadiene	284° C	20
Nornicotine	270° C	8
Dotriacontane	467° C	30
Solanesol	(M.P.) 41.5° C	3

*BASED ON CONTENT IN TOBACCO CONSUMED

Table 2.   Some Types of Compounds Found in Tobacco Smoke.

Aliphatic Hydrocarbons	Naphthols
Aromatic Hydrocarbons	Nitro Compounds
Carbazoles	Nitriles
Many Esters	Piperidines
Furans	Pyrazines
Fluorenes	Pyrrolidines
Indans	Toluidines
Indoles	

smoke.   Certain low molecular decomposable substances like amino acids have also been found.   The presence of all these has been explained by the hypothesis that under thermal stress the tobacco cells "explode" and eject into the stream involatile solids. One must also remember that, especially with non-filter cigarets, tobacco particles are often swept into the smoke collection devices during the puff.   In both of these ways unexpected involatile substances could show up in a smoke condensate analysis.   Starch, proteins, pectins, chlorophyll and the like have not been identified in smoke and would, in general, not be expected there.

Finally, there are also in smoke the products of pyrolysis, oxidation, and high temperature synthesis.   Some examples of these are shown in Table 2.

These and other "new" substances which are the result of reactions occurring during smoking represent a large part of the 900 identified constituents in smoke.   The chemistry of their

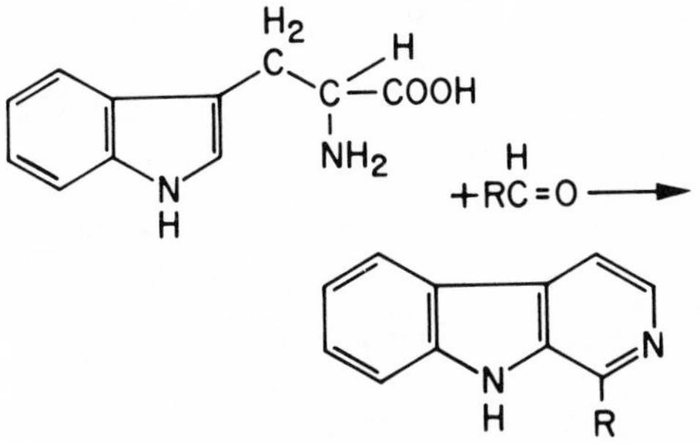

R=H, NORHARMANE;   R=CH$_3$, HARMANE

Figure 2.   Formation of Harmanes from Tryptophan and Aldehydes.

Table 3.   Tryptophan to Harmanes Conversion in the Cigaret.[34] [35]

A. Radiotracer Experiment

169 nci ^{14}C-Tryptophan added per cigaret

	YIELD	
	NORHARMANE	HARMANE
Activity in Smoke	0.80 nci	0.26 nci
% Conversion	0.47	0.15

B. Tobacco Loading Experiment

Control Yield	11.2 μg	4.3 μg
Added 820/μg Tryptophan/Cigaret		
Observed	15.7 μg	5.9 μg
Increase	4.5	1.6
Predicted	3.9	1.2

formation is naturally a subject of great interest to the tobacco chemist. Accordingly, we find an increasing number of investigators looking into the question of what and how tobacco constituents are precursors of chemicals formed during the burning process.

Three different approaches have been employed. In the first, tobacco has been loaded or spiked with the suspected precursor and smoked. The smoke is then analyzed for increases in amounts of the suspected products. Here are some examples of this technique (Fig. 2, Table 3, Fig. 3):

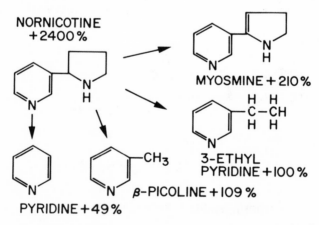

Figure 3.   Effect of Nornicotine Addition.[11]

The first example (Fig. 2) involves the conversion of the amino acid tryptophan to the carbolines harmane and norharmane by reaction with an aldehyde.[34][35] Tryptophan is present in tobacco either as an amino acid or as a component of tobacco protein. In a second experiment (Table 3) 820 μg/cigt of tryptophan added to the tobacco gave an increase in norharmane and harmane yields about as expected from the radiotracer data.

Another example (Fig. 3) involves the addition of nornicotine, the demethylated nicotine analog.[11] A 24-fold increase of nor- nicotine produced the indicated increases in the nitrogen bases which might be expected as breakdown products from nornicotine. One of the problems in this type of experiment is that the added precursor is probably not located in the tobacco in the same way as the naturally occurring one. As the thermal stress develops the evaporation and/or pyrolysis may not be the same for the added component as for the natural one. The breakdown products might also be different. Another difficulty with this type of experiment is that substantial loading of the tobacco with the precursor in question may change the burning properties of the tobacco.

A second approach to the problem of identifying precursors has been to pyrolyze in a hot tube or furnace individual precursor candidates and to analyze the pyrolysis products. This technique is useful for indicating what compounds might be looked for in the more sophisticated radiotracer experiments; or, conversely, it may also be used to identify many precursors for a given smoke con- stituent.

For example (Table 4), many substances when pyrolyzed yield benzo(a)pyrene.[24] [12] [36]

Table 4. Benzo(a)pyrene in Pyrolysis Products.[24]

PRECURSORS	μg B(a)P FROM 100 G OF STARTING MATERIAL
Triethylene Glycol	3
Glycerol	6
Sorbitol	13
Starch	7
Cellulose	8
Hydroxyethyl Cellulose	34
Gum Arabic	32
Agar – Agar	47

Obviously, to pinpoint a specific precursor in tobacco would be a futile exercise. A similar situation exists with the occurrence of phenol in smoke (Table 5).

Bell and co-workers (4) pyrolyzed various carbohydrates in air and nitrogen to show that many form phenol at elevated temperatures. It is interesting to note that higher phenol yields are obtained in the absence of air with its competing oxidation reactions (Table 5).

The more extensive analysis of pyrolysis products from malic acid is illustrated in Table 6.

Here only the presence or absence of products are indicated. It is interesting to observe that at higher temperatures the lower weight oxygenated products give way to polynuclear hydrocarbons. Thus, we obtain by this method not only scope of the pyrolysis products, but also clues regarding their temperatures of formation and thermal stability.

Pyrolysis experiments do give some information about thermal conversions of tobacco constituents into smoke chemicals. The problem arises in the interpretations of the results. Pyrolysis conditions only approximate the burning cigaret and make no allowance for the presence of other tobacco components.

This leads us to the third general technique of studying precursors -- that employing labeled compounds. Here we may use either radioactive or mass isotopes. The radioactively labeled compounds can be added to tobacco in extremely small quantities without interfering with the normal burning characteristics of the tobacco.

Uniformly labeled radioactive ^{14}C-glucose and ^{14}C-sucrose were applied to cigaret tobacco by Gager and co-workers (8) (9) and by Thornton and Valentine. (45) The cigarets were smoked with the results shown in Table 7.

Similar results were obtained with sucrose. Note that over half of the combustion products appear in sidestream smoke and that a wide variety of radioactive oxidation and pyrolysis products were identified. Of particular interest is the presence of radioactive acetonitrile which would not have been found in a pure compound pyrolysis experiment. It must have been formed by reaction with nitrogenous compounds in the tobacco. Another point is of interest in connection with our earlier remarks about entrainment in the smoke stream. Gager and co-workers report 0.5% of the original radioactive sugar in the mainstream particulate smoke.

So far we have considered the general field of tobacco smoke

Table 5.　Pyrolysis of Carbohydrates to Phenol.[4]

PRECURSOR	100 (% CONVERSION)	
	IN AIR	IN N_2
Glucose	5.0	9.0
Sucrose	6.7	8.2
Starch	0.2	10.0
Cellulose	2.4	4.0
Pectin	2.6	3.9
Rutin	0.8	2.8

MAX. TEMP. 685°C; GAS FLOW 100 ml/min.

Table 6.　Products from Pyrolysis of Malic Acid Observed at Various Temperatures.[10]

	500°	600°	700°	800°	900°C
Benzene	+	+	+	+	+
Toluene	+	+	+	+	+
Indene	+	+	+	+	+
Biphenyl			+	+	+
Naphthalene			+	+	+
Ethylbenzene } m-,p-Xylene			+	+	+
Styrene } o-Styrene			+	+	+
Dimethylnaphthalene					+
Acenaphthylene				+	+
Anthracene (Phenanthrene)				+	+
Fluorene				+	+
Pyrene					+
Chrysene					+
Phenol	+	+	+		
m-Cresol	+	+	+		
p-Cresol	+	+	+	+	
o-Cresol	+	+	+	+	
2,5-Xylenol		+	+	+	
Fumaric Acid	+	+	+		
Succinic Acid	+	+			+

Table 7.   Distribution of Products from ^{14}C-Glucose in Tobacco.[8][9]

	% TOTAL ACTIVITY
Butt	4I
Sidestream Smoke	5I
Mainstream Smoke	
Particulate	I.7
CO_2	2.7
Other Gases*	2.0

* Identified Compounds Include

Acetaldehyde	2-Butanone
Acetone	2-Buten-2-One
Acetonitrile	
Acrolein	2,3-Butanedione
Benzene	Crotonaldehyde
Furan	2,5-Dimethylfuran
2-Methylfuran	Propionaldehyde

chemistry in a more or less traditional way. To go further into
the subject it will be useful at this point to consider the reaction conditions in the burning zone in considerably greater detail
than heretofore. We will for the sake of simplicity limit ourselves
to the cigaret (Fig. 4), although much of what has already been said
applies to pipe and cigar smoking as well.

Smoke is produced by the smoker puffing through the cigaret
and by smoldering between puffs. The two smokes, identified as
mainstream and sidestream smokes, are not the same. Both streams
are composed of particulate phase and a gas phase, generally defined
as that portion not removed by the conventional Cambridge filter.

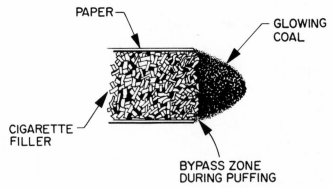

Fig. 4. Longitudinal Section of the Burning Cigaret.

After the cigaret is lit and allowed to smolder a bit, a burning cone appears at the ignited end (Fig. 4). This cone has a higher density of material than the tobacco behind it, produced by coking and swelling of the tobacco shreds and by the recondensation of evaporation and pyrolysis products induced by the smoldering process itself. One manifestation of this dense cone is an increase in the resistance to draw of the overall cigaret, a reduced permeability to air. At the same time the porosity and burning rate of the cigaret paper tends to keep the periphery of the cone unobstructed. In fact, the paper burns away ahead of the tobacco to a point where further burning is halted by the quenching effect of air flow to the burning cone from behind.

Now comes the puff, drawing air past the burning coal around the periphery of the cigaret. This flow provides more oxygen to this area and increases its maximum temperature, much like a blacksmith blowing air on his charcoal to increase the temperature of his forge. The hot gases from this zone, drawn into the cigaret, again swell the tobacco filaments behind the cone and reduce the airflow. The combustion zone is still further bypassed, with the result that for very strong puffs, the volume of cigaret consumed is actually less than for weaker puffs. [14] During the puff, then, the shape of the burning cone is distorted from its natural smoldering shape. This shape is gradually restored during the interval between puffs.

In the swelling of and condensing on the tobacco behind the burning zone during the puff, the otherwise normal flow of oxygen to the backside of the cone is reduced. [7] [14] [41] This "choking" off of the oxygen persists for a time after the puff with the result that the rate of tobacco consumption is actually lower than for the natural smoldering process. This fact is borne out by the observation of an increase in smoldering temperature some time after the puff, usually about forty seconds into the interval. The normal oxygen supply is now restored.

The maximum temperature for the burning tobacco, periphery or coal will fluctuate as shown in Fig. 5 for puffs and between puffs. [26]

Oxygen availability to the burning zones will likewise vary. Both temperature and oxygen will then depend on the velocity of the air stream during puffing, the duration of the puff, the interval between puffs, and the density of the tobacco rod and coal as they affect heat transfer and radiation and convection losses. It is clear from these considerations that the chemistry of the smoke will vary (1) between mainstream and sidestream smoke for a given puff condition and (2) among various puff conditions.

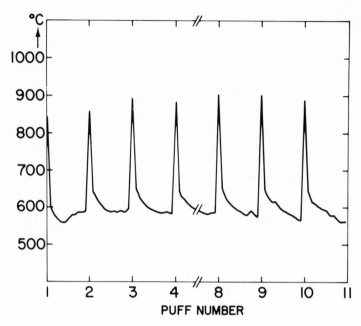

Fig. 5. Temperature Scan of Burning Region Showing Maximum Values.[26]

We need, therefore, a standard puff condition for chemical studies of the cigaret. As most of you know, this has been established for the Federal Trade Commission test on tar and nicotine. (33) The test involves a 35 ml volume puff of two seconds duration once a minute. This test condition is an arbitrary choice based on a consideration of how the "average" smoker puffs his cigaret.

It should be stressed that smokers may vary widely from this standard smoking pattern. (1) (22) The observed ranges are shown in Table 8 for a number of smoking parameters.

Note that flow rates for air drawn into the cigaret during the puff by different people exhibit a 1 to 10 ratio for the lowest

Table 8.   How Smokers Puff Cigarets.

PUFF	LOWEST	STANDARD	HIGHEST
Volume ml	17.0	35.0	73.0
Duration Sec.	0.9	2.0	3.2
Rate ml / Sec.	5.6	17.5	81.0
Interval Sec.	22.0	60.0	72.0

Table 9. Ratios of Smoke Constituents in Sidestream and Mainstream
(Based on Yields per Gram of Tobacco Consumed).

CONSTITUENT	SS/MS
Acetic Acid	0.42
Acetamide	0.68
HCN	0.03
Acetonitrile	2.3
Aldehydes (Total)	2.2
3,4-Benzo(a)Pyrene	2.1
Benzene	1.9
Toluene	3.5
Phenol	2.1
Nicotine	2.5
Myosmine	7.0
Pyrrole	7.1
Pyridine	9.5
3-Vinylpyridine	16.0

to the highest. At these extremes we would expect the smoke chemistry to vary considerably. The smoker taking a light short puff
may be not only obtaining less smoke per puff from the cigaret, but
also smoke that is different chemically from that obtained by the
smoker taking a long hard puff.

It is interesting to compare the composition of mainstream (MS)
and sidestream (SS) smoke at the standard smoking condition (Table 9).

Table 9 shows the ratio of constituents in sidestream to
mainstream smoke for a number of components. It is based on
observed yields per gram of tobacco consumed. Note that the first
three constituents exhibit greater yields during mainstream smoking;
the next seven have approximately twice the mainstream yield in
sidestream smoking; and that the next four nitrogen ring compounds
have about ten times the yield under the sidestream smoking conditions.

Johnson and Kang (18) (19) have made a study of the mechanisms
of HCN formation from the pyrolysis of nitrogen compounds. They
observed in some cases almost total nitrogen conversion to HCN at
1000°C with rapid decrease in yields as pyrolysis temperatures are
decreased to 800°C.

Their observation fits in with the temperature differential
between mainstream and sidestream smoking noted earlier. In contrast with the higher HCN yield of mainstream smoking is the lower
yield of the nitrogen bases in the mainstream. Many of these are
precursors of HCN. We have then the general picture that at the

Table 10. Ratios of Gaseous Smoke Products in Sidestream and
Mainstream.

CONSTITUENT	SS/MS
Ammonia	98.0
Carbon Dioxide	3.8 — 10.0
Carbon Monoxide	1.3 — 3.0

CO_2/CO RATIOS	
Mainstream	2.2 — 6.0
Sidestream	10.0 — 13.0

lower sidestream smoldering temperatures nitrogen compounds are
only partially decomposed and evaporated into the smoke; whereas,
under hotter mainstream smoking conditions, more complete pyrolysis
to HCN takes place.

The pyrolysis, oxidation, and pyrosynthesis products in
Table 9 are all more prevalent in the sidestream smoke than in the
mainstream. Please be reminded that, in calculating these ratios,
we have already corrected for the fact that in the standard puffing
sequence twice as much tobacco is consumed during the smoldering
period as in the puffing period. Smoldering seems to involve slow,
low temperature oxidation and pyrolysis in a relatively oxygen-
starved atmosphere. Puffing, on the other hand, involves the rapid
flow of air past the periphery of the burning coal. [3]   Oxygen
is available for rapid combustion of the tobacco at high temperatures. But the hot gases flow quickly to heat the tobacco behind
the coal, sweeping with them not so much pyrolysis products as
volatiles distilled from the tobacco.

In examining this picture, it is instructive to consider the
major gases.

Note in Table 10 that ammonia,[48] carbon dioxide, and even
carbon monoxide are far more prevalent in sidestream smoke. Note
also that the CO_2/CO ratio is also much higher in the sidestream.[20]
Clearly much more oxidation is taking place in sidestream than in
mainstream smoking.

The very high preponderance of ammonia in sidestream and the
virtual absence in mainstream smoke could be due to a number of
factors. Firstly, ammonia is extremely sensitive to high temperature dissociation. [25] The reaction to form ammonia from nitrogen
and hydrogen is exothermic so that ammonia formation is favored at
low temperatures (Fig. 6).

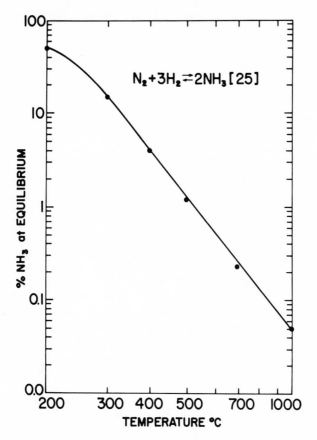

Fig. 6.   Ammonia Decomposition Equilibria at Elevated Temperatures.

        The values of ammonia concentration at equilibrium indicate
that less and less will be in equilibrium with nitrogen and hydrogen at the higher temperature.   A further point in this connection
is the existence of free hydrogen in mainstream smoke gases.
(30) (23)

        Secondly, ammonia will react with the free acids and other pro-
ducts distilled from the tobacco.   We have already noted the preponderance of acetic acid and the secondary ammonia reaction product, acetamide, in mainstream smoke.

        Thirdly, some ammonia formed in mainstream smoking may be
absorbed by the tobacco through which the mainstream is passing.
Ammonia readily reacts with the sugars in tobacco to form a variety
of reaction products.   It also has been shown to displace bound
water in cellulose. (46)   There are thus several mechanisms whereby

ammonia, if present in mainstream smoke, would be removed before
the smoke issues from the cigaret.

Finally, I would like to summarize several recent radiotracer
studies which support the cigaret model as described. In these
experiments, C-14 labeled compounds have been incorporated into the
cigaret tobacco. The distributions (%) of radioactivity in the
various smokes were found to be as shown in Table 11. Let us consider each of these compounds briefly.

Menthol [17] is the most volatile compound studied. It transfers as menthol to a great extent in the mainstream particulate
without significant pyrolysis or degradation into the mainstream
gas. Some is consumed in sidestream burning to form gaseous products and/or pyrolysis products in the sidestream particulate.

Dotriacontane [16] is a C-32 hydrocarbon of great stability.
It transfers to the mainstream particulate with 95% of the activity
identifiable as unchanged dotriacontane. Again there are negligible
mainstream gaseous byproducts, and relatively small amounts of combustion products in the sidestream gases. Most of the radioactivity
in the sidestream is in the particulate, presumably unchanged hydrocarbon or cracking products.

Nicotine [15] also transfers well to the mainstream particulate
as nicotine, but is less stable than dotriacontane. It breaks
down to other nitrogen bases in sidestream smoking. These appear
mainly in the particulate of the sidestream.

Table 11. Radioactivity Distribution (%) in Smoke from Tobacco
Labeled with ^{14}C-Compounds (Based on Tobacco Consumed).

COMPOUND	MAINSTREAM		SIDESTREAM	
	GAS	PARTICULATE	GAS	PARTICULATE
Menthol	1.4	38.0	29.0	32.0
Dotriacontane	1.9	28.1	15.0	55.0
Anthracene ^{14}C(U)	0.3	30.0	2.4	62.0
Benzo(a)Pyrene ^{14}C(7:10)	0.0	29.5	2.4	72.0
Nicotine	3.0	21.0	21.0	55.0
Glucose ^{14}C(U)	7.3	2.7	73.0	5.9
Sucrose ^{14}C(U)	9.4	4.8	82.5	6.1
Oxalic Acid	19.4	5.0	73.9	0.8
Citric Acid	12.5	3.4	76.8	4.4
Maleic Anhyd. ^{14}C(1:4)	22.0	3.1	55.0	3.7

Glucose and Sucrose [8] are nonvolatile sugars which decompose readily on heating. They do not transfer intact to any extent to either mainstream or sidestream smoke. Most of the gas phase activity in both streams is in carbon dioxide, although some pyrolysis and oxidation products are also found.

Oxalic and Citric Acids [29] both decarboxylate at elevated temperatures; hence most of the activity in both main- and sidestream gases is found as carbon dioxide and carbon monoxide. The particulate in mainstream contains a very small amount of original acid, the remainder probably being pyrolysis decomposition products.

These experiments demonstrate that in the mainstream smoking the more stable volatile compounds in the tobacco transfer intact almost quantitatively to the smoke stream with very little pyrolysis or oxidation taking place. This transfer must be occurring from the heated tobacco well behind the burning zone. Nonvolatile substances remain to be consumed in sidestream smoking. Only those which are very susceptible to thermal degradation or those which require a higher temperature [27] yield reaction products which appear in the mainstream gases. Finally, there is clearly some contamination of this mainstream smoke by combustion products from the smoldering cone, products which otherwise would only be in sidestream smoke. [38]

What happens to the mainstream components as they pass through the tobacco? Some of the gaseous substances probably pass through unhindered. Others may diffuse out through the porous paper. Such effects have been shown for hydrogen, carbon monoxide and carbon dioxide. [32] [44] Still others will react with tobacco compounds, or be absorbed by the tobacco to provide some buildup in the butt as the cigaret gets shorter and shorter. [39] [47] Water is one such component. It is absorbed and increases the moisture content of the remaining unburned tobacco. It is also a part of the particulate phase, "partitioning" between it and the gaseous phase. [49]

Some components absorbed by the tobacco and some volatile tobacco constituents are probably "eluted" off the tobacco by the mainstream just as they have been shown to be eluted from the filter. [6] Finally, we have those less volatile substances which make up the particulate mainstream smoke. These are also passed through, relatively unhindered. The tobacco is a very poor particulate filter.

This last point has proved to be the source of considerable discussion after previous presentations of the subject. So, before the critics jump out of their seats, I want to present one more table based on the radiotracer experiments I have just described (Table 12).

Table 12. Decomposition on Tobacco in the Unburned Portion of the Cigaret from Radioactive Tracer Experiments.

	% TOTAL ACTIVITY		
COMPOUND	OBSERVED	EXPECTED FROM BUTT PORTION	EXCESS
Menthol	26.9	28.5	−1.6
Dotriacontane	30.0	29.9	−−
Nicotine	29.8	29.9	−−
Glucose	41.0	37.3	+3.7
Sucrose	40.6	37.3	+3.3
Oxalic Acid	0.22	0.0	+0.22
Citric Acid	0.71	0.0	+0.71

Here I have tabulated the measured fraction of total activity of the unburned portion or butt left after the cigaret has been smoked, and compared it with the expected activity of the butt, based on the fraction of the cigaret not consumed in the smoking. The agreement is almost too good. Menthol, because of its volatility, is actually swept off by the mainstream so that the residual activity is depleted in the butt end. Glucose and sucrose show some slight deposition of radioactivity. In the oxalic and citric acid experiments, only the front 45 mm of the cigarets were impregnated and 50 mm were consumed in smoking so no radioactivity was expected in the butt. It seems that if there is buildup in the unburned to-bacco of smoke products, it must come from other mechanism than by deposition of partₜculate mainstream smoke.

In summary, I have tried to indicate the gradual transition from emphasis on analysis of smoke to the more interesting aspects of smoke formation and the chemical processes involved. I have described to the limited extent of our present knowledge the con-ditions which seem to prevail during both puffing and smoldering of the cigaret. I have illustrated how the use of labeled com-pounds provides us with a more sophisticated tool for studying the chemistry of the burning tobacco. I have probably raised more questions than I have answered, but this is the challenge of research.

References

1. Adams, P.I. (1966) Presented at the 20th Tobacco Chemists'
    Research Conference. #31.
2. Adams, P.I. (1968) Tob.Sc.XII, 144.
3. Baxter, J.E. and Hobbs, M.E. (1967) Tob.Sc.XI, 65.
4. Bell, J.H., Saunders, A.O. and Spears, A.W. (1966)Tob.Sc.X, 138.
5. Bruckner, H. (1936)Die Biochemie des Tabaks Verlag Paul Parey,
    Berlin.
6. Curran, J.G. and Miller, E.G., Jr. (1969) Beitrage zur
    Tabakforschung 5, 64.
7. Egerton, A., Gugan, K. and Weinberg, F.J. (1963) Combustion &
    Flame 7 #1, 63.
8. Gager, F.L.,Jr., Nedlock, J.W. and Martin, W.J. (1971 a) Carbohyd.
    Res. 17, 327.
9. Gager, F.L.,Jr., Nedlock, J.W. and Martin, W.J. (1971 b)Carbohyd.
    Res. 17, 335.
10. Geisinger, K.R., Jones, T.C. and Schmeltz, I. (1970)Tob.Sc.XIV,89.
11. Glock, E. and Wright, M.P. (1962)Presented at the 16th Tobacco
    Chemists' Research Conference, #22.
12. Grimmer, G., Glaser, A. and Wilhelm, G. (1966) Beitrage zur
    Tabakforschung 3, 415.
13. Grob, K. and Vollmin, J.A. (1969) Beitrage zur Tabakforschung 5,
    52.
14. Gugan, K. (1966) Combustion & Flame 10, 161.
15. Jenkins, R.W.,Jr., (1971) Philip Morris Research Center. Private
    Communication.
16. Jenkins, R.W.,Jr., Newman, R.H., Carpenter, R.D. and Osdene, T.S.
    (1970 a) Beitrage zur Tabakforschung 5, 295.
17. Jenkins, R.W., Jr., Newman R.H., and Chavis, M.K. (1970 b)
    Beitrage zur Tabakforschung 5, 299.
18. Johnson, W.R., Kang, J.C. and Wakeham, H. (1970)Presented at
    the 5th International Tobacco Scientific Congress, Hamburg.
    #B403.
19. Johnson, W.R. and Kang, J.C. (1971) J.Org.Chem. 36, 189.
20. Johnson W.R., Hale, R.W., and Nedlock, W.J. (1971) Philip Moriis
    Research Center. Private Communication.
21. Johnstone, R.A.W. and Plimmer, J.R. (1959)Chem.Rev. 59, 885.
22. Keith, C.H. (1962) Presented at the 16th Tobacco Chemists'
    Research Conference. #24.
23. Keith, C.H. and Tesh, P.G. (1965) Tob.Sc.IX, 61.
24. Kroller, E. (1965) Deut.Lebensm,Rundschau, 16.
25. Larson, A.T. and Dodge, R.L. (1923)J.Am.Chem.Soc. 45, 2928.
26. Lendvay, A.T. (1971) Philip Morris Research Center. Private
    Communication.
27. Morrell, F.A. and Varsel, C. (1966) Tob.Sc.X, 45.
28. Neurath, G., Ehmke, H., and Schneemann, H. (1966) Beitrage zur
    Tabakforschung 3, 351.

29. Newell, M.P. and Best, F.W. (1968) Presented at the 22nd Tobacco Chemists' Research Conference. #25.
30. Newsome, J.R. and Keith, C.H. (1965) Tob.Sc.IX, 65.
31. Okada, T., Ishibashi, H., Shimada, Y. and Obi, Y. (1968) Tob.Sc.XII, 105.
32. Owen, W.C. and Reynolds, M.L. (1967) Tob.Sc.XI, 14.
33. Pillsbury, H.C., Bright, C.C., O'Connor, K.J. and Irish, F.W. (1969) Journal of the AOAC Vol. 52 No. 3.
34. Poindexter, E.H.,Jr., and Carpenter, R.D. (1962) Phytochemistry Vol. 1, 215.
35. Poindexter, E.H., Jr., Bavley, A. and Wakeham, H. (1963) Proc.3rd World Tobacco Sci. Congr. 550.
36. Robb, E.W., Johnson, W.R., Westbrook, J.J. and Seligman, R.B. (1966) Beitrage zur Tabakforschung 3, 597.
37. Schmeltz, I., Hickey, L.C. and Schlotzhauer, W.S. (1967) Tob. Sci.XI, 52.
38. Seehofer, F. and Schulz, W. (1965 a)Beitrage zur Tabakforschung 3, 151.
39. Seehofer, F., Hansen, D. and Schroder, R.(1965 b) Beitrage zur Tabakforschung 3, 135.
40. Shmuk, A.A. (1953) The Chemistry and Technology of Tobacco, Pishchepromizdat, Moscow. (Translation available from Office of Technical Services, U.S. Dept. of Commerce, Wash. 25, D.C.)
41. Somasundaran, P. (1970) Tob.Sc.XIV, 156.
42. Stedman, R.L. (1968) Chem.Rev. 68, 153.
43. Stedman, R.L., Benedict, R.C., Dymicky, M. and Bailey, D.G. (1969)Beitrage zur Tabakforschung 5, 97.
44. Terrell, J.H. and Schmeltz, I. (1970) Tob.Sc.XIV, 82.
45. Thornton, R.E. and Valentine, C. (1968)Beitrage zur Tabak-forschung 4, 287.
46. Walker, G.B.,Jr. (1971) Philip Morris Research Center. Private Communication.
47. Waltz, P. and Hausermann, M. (1965) Beitrage zur Tabakforschung 3, 169.
48. Williams, J.F. and Hunt, G.F. (1967) Presented at the 21st Tobacco Chemists' Research Conference. #24.
49. Williamson, J.T. and Allman, D.R. (1966) Beitrage zur Tabak-forschung 3, 591.

HIGH MOLECULAR WEIGHT MATERIALS OF TOBACCO

O. T. Chortyk

Russell Research Center, SEMNRD, USDA

Athens, Georgia   30604

The present report constitutes a review of our current knowledge on the high molecular weight constituents of tobacco. Because
of the vast literature on the subject, only selected highlights will
be discussed. Generally high molecular weight substances of leaf
may be divided into three groups: proteins, polysaccharides, and
polymeric polyphenols. The polysaccharides include cellulose,
starch, and the pectins. The polyphenols include lignin and the
brown pigments. The proteins, polysaccharides, and polymeric polyphenols together comprise from 30 to 40% of tobacco leaf, with the
polysaccharides being the most abundant group. Due to their extensive occurrence, a thorough knowledge of their chemical and physical
properties is essential to our understanding of leaf composition and
the effect upon smoke quality and smoke-health hazards. Certainly,
the pyrolytic products of the polysaccharides will contribute significantly to smoke composition and will moreover affect the organ-
oleptic characteristics of the smoke. However, this topic is the
subject of other presentations and will not be discussed here. In
the present discussion the high molecular weight substances will be
described in terms of their extraction from tobacco leaf and subsequent characterization through physical and chemical methods with
established or proposed structures being presented where possible.

The first group that will be considered are the proteins.
Various enzymatic activities associated with them will only be mentioned where pertinent and all aberrations induced by tobacco
mosaic virus (T.M.V.) will not be included. One of the methods of
protein extraction is shown in Figure 1. Proteins were extracted
from green or cured leaf by homogenizing the leaf with an appropriate buffer system that contained the antioxidants: ascorbic
acid and cysteine. The resulting mixture was centrifuged and the

21

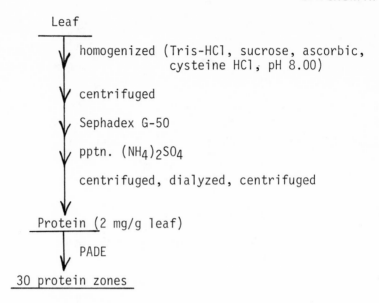

Fig. 1. Protein Extraction according to VanLoon and Kammen (1).

protein separated from yellow pigments and low molecular weight
materials by gel filtration and concentrated by ammonium sulfate
precipitation. Subsequent dialysis and centrifugation yielded
pure protein that was separated by polyacrylamide disc electro-
phoresis (PADE) into about 30 protein zones.

Another extraction is shown in Figure 2. In this method, the
protein is separated by centrifugation. Centrifugation at 5000 g
yields the chloroplast Fraction I residue, while the supernatant
solution is subdivided by centrifugation at 100,000 g into chloroplast Fraction II and the supernatant Fraction III. Each fraction
is then purified by precipitation with 20% trichloroacetic acid
(TCA), dissolution with 0.1N sodium hydroxide, subsequent precipitation with TCA, and washing with 85% acetone.

Since protein occurs in both green and cured tobacco leaf, it
is interesting to observe the changes in protein associated with
this transition from green to cured leaf, and much work has been
done in this area. Results from such a study (i.e., variation as a
function of aging) are shown in Table 1. As is apparent from the
data, protein content decreases rapidly with aging. This decrease
is accompanied by an increase in protease activity and a resulting
increase in free amino acids and amides. Results of another study
in which protein changes associated with curing were measured are

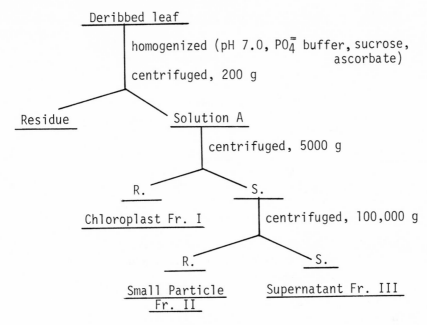

Purefication: TCA, NaOH, TCA, Acetone

Fig. 2. Protein Extraction according to Kawashima (2).

Table 1. Protein Variation with Aging (in mg/15 g fresh leaf)

	Weeks after transplantation		
	2	4	7
Solution A	176	179	54
Fr I	63	52	16
Fr II	12	15	9
S III	101	112	29

green leaf $\xrightarrow[\text{decrease}]{\text{proteins}}$ cured leaf

free amino acids, amides increase ➤

protease activity increases ➤

Table 2.   Protein Change during Curing

Hours	0	20	68
Dry wt., g	2.17	1.97	1.75
Total protein, mg	144	115	66
Solution A, mg	103	90	40
Fr I, mg	33	26	8
Fr II, mg	10	8	2
Fr III, mg	59	56	30

Fr I glutamic and basic amino acids increased;
    proline, glycine, leucine decreased

Fr III basic amino acids decreased; proline,
    aspartic, serine increased

given in Table 2. Here protein fractions were hydrolyzed and the
resulting amino acids were determined (3). It was found that the
proportions of various amino acids changed, indicating perhaps that
certain proteins within a fraction were preferentially destroyed.

The changes in soluble leaf proteins on curing in four different tobaccos were studied by PADE (4), and it was concluded
that the effects of curing on the proteins were the same for all
tobaccos. Generally, 13 protein bands, found initially, decreased
to 4 bands after about 17 days of air-curing. Three of the 4
common bands appeared to be glycoproteins. As with other studies,
it was observed that Fraction I protein disappeared after about a
week of curing. Kawashima has examined these changes by chromato-
graphing the total protein, obtained by 100,000 x g centrifugation,
on Sephadex G-200 (5). The chromatograms of the "before-curing
proteins" showed seven distinct bands, as determined by UV-absor-
bance and protein-nitrogen determinations. The "after-curing
proteins" showed only about three bands. Fraction I protein, which
was the largest protein in the extract and constituted more than
half of the soluble protein before curing, was not found in the
proteins after curing. However, proteins contained in the last
three bands, which had smaller molecular weights, increased about
threefold on curing. Thus overall, it is concluded that high
molecular weight proteins are converted to low molecular weight
proteins on curing.

Of predominant interest has been the high molecular weight protein band, designated as "Fraction I protein". This material occurs in most plants and has been extensively studied in such plants as tobacco, clover, spinach and cabbage (6). Some of its main characteristics are presented in Table 3. When extracts of plant leaves are prepared at pH 6.5 and the chloroplasts are disrupted, the analytical centrifugation of such extracts will usually produce a schlieren pattern where about 50% of the area of the pattern may consist of a simple protein component with a sedimentation constant of about 18 S, which is designated Fraction I protein. Based on the sedimentation constant, the molecular weight is about 550,000 and it is found to be a compact and isodiametric particle as seen by electron microscopy. Differences in the amino acid compositions have shown that the primary structure of Fraction I protein from different plants is not identical. Purified Fraction I protein from different plants has been shown to be homogeneous in regard to their capacity to combine with antibody.

Efforts to isolate pure 3-phospho-d-glycerate carboxylaze, known as RuDP carboxylase, have yielded a protein that is in all respects identical with Fraction I protein. RuDP carboxylase catalyzes the carboxylation and cleavage of D-ribulose 1,5 diphosphate into 2 moles of 3-phosphoglyceric acid. Based on the work of numerous investigators, it appears that RuDP carboxylase is inseparable from Fraction I protein. Both Fraction I protein and the enzyme activity are located almost exclusively in the chloroplasts, and both exhibit the same antigenic activity. Thus, recovery of the precipitated Fraction I protein-antibody complex demonstrated that the complex was capable of forming 3-phosphoglyceric acid from D-ribulose 1,5 diphosphate and carbon dioxide. Thus, the purified product, whether obtained as the protein or enzyme, meets the ordinary criteria for enzymatic, electrophoretical and ultracentrifugal homogeneity.

Table 3.   Protein Characteristics

1. Molecular weight - 550,000
2. Electron microscopy - compact, isodiametric
3. Carbohydrates ?
4. Amino acid composition varies
5. Active antigen, no bimodality
6. Two subunits
7. Identity with RuDP carboxylaze activity

In order to study the nature of the subunits of the protein, Kawashima dissociated it by sodium dodecyl sulfonate and by alkali at different pH values and then performed gel filtration (7). The dissociation was pH dependent. Thus, at pH 11, it was not completely dissociated while at pH 11.2 three components could be resolved by gel filtration on Sephadex G-200: 500,000, 52,000 and 24,000. Amino acid analyses showed that the two higher molecular weight components were identical and that the 500,000 material consists of reaggregated 52,000 protein. The 24,000 protein was different in composition and was believed to consist of two 12,000 entities. In conclusion, it seems that similar work on the other 30 or more proteins of tobacco leaf still remains to be accomplished and their exact composition and activities still remain to be determined.

The next group of high molecular weight substances to be discussed are the polysaccharides, which include cellulose, starch and the pectins. The chemistry of cellulose has been well explored especially for various woods. It is accepted that cellulose is composed of long chains of β-d-glucopyranose units with molecular weights ranging from 400,000 to about 1,800,000. Due to its large molecular weight, it is relatively inert to most organic solvents and weak acid or base treatments. Such treatments usually remove all plant constituents, leaving cellulose as a residue. Thus, from tobacco, cellulose remains after extracting the tobacco with water, ethanol, acetone, benzene, 6% alkali, 1% sodium dithionate, 50% acetic acid and water. Christy and Samfield (8) studied the extraction, distribution and degree of polymerization (DP) of cellulose in various tobacco types. As expected, the more rigid tobacco stems contained greater percentages of cellulose (16.5 to 24.7%) than the lamina (7.2 to 10.2%). DP is defined as the average number of anhydroglucose units in the chain molecule of cellulose. Stems had DP values of 1600 to 1800, while the lamina values ranged from 1340 to 1600.

The isolation of a related group of carbohydrates is detailed in Figure 3. These are the hemicelluloses. Methanol-water (M-W) and weak sodium hydroxide and ammonium oxalate remove low molecular weight materials and the pectins. Lignin is subsequently removed by the sodium chlorite-acetic acid method or by reflux with 1% sodium hydroxide in 50% ethanol. The hemicelluloses are then extracted consecutively with 5, 10 and 20% potassium hydroxide solutions giving the fractions and the yields shown. The composition and molecular weights of the hemicelluloses are shown in Table 4 (9).

As contrasted to cellulose, the major carbohydrate unit in hemicellulose is xylose, with smaller amounts of glucose and arabinose. A hemicellulose was also obtained which contained only xylose. This indicated that the hemicelluloses consist of

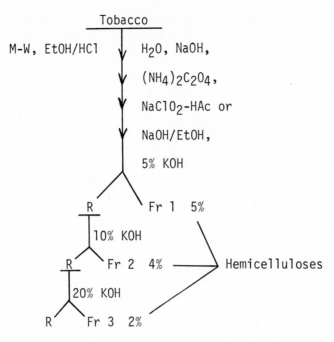

Fig. 3.   Isolation of Hemicelluloses from Tobacco.

Table 4.   Composition and Molecular Weights of Hemicellulose

	Xylose	Glu	Arab
Composition			
Bright stems 5% KOH	74	23.5	2.5
10% KOH	86.6	9.2	2.2

Molecular Weights: Fr 1 Components

Maryland stems	95,000	8,000
Bright stems	75,000	11,000
Havana leaf	110,000	5,500

a xylose backbone with other carbohydrate moieties attached as
side chains. Gel-permeation chromatography indicated that Fraction
1 hemicellulose of stems was composed of a high and low molecular
weight material as shown.

The next carbohydrate polymer that will be considered is
starch. As a result of numerous hydrolytic experiments, it has
been concluded that starch is composed of chains of alpha-d-
glucopyranose units. The two large components are amylose and
the less soluble, more abundant amylopectin. Starch is usually
extracted from tobacco with perchloric acid although the other
reagents such as ammonium oxalate, chloral hydrate and sodium
hydroxide have also been used. Starch is determined as the blue
iodine complex (10), and using such a method the content of green
leaf has been shown to be as high as 20% while that of cured leaf
is about 2-3%.

The final group of polysaccharides are the pectic substances,
defined in Table 5. The pectins are polymers of polygalacturonic
acid and are distinguished from each other on the basis of their
water solubility and the extent of methylation of the acid groups.
Thus, pectin, although greatly esterified, is the smallest, and
most soluble, while protopectin, regarded as the parent material,
is the largest and most insoluble material. The total pectic
substances are obtained from tobacco, after a preliminary extraction with 95% ethanol, by treatment with water, dilute acid,
dilute base, ammonium oxalate or ammonium citrate (11). Using
ammonium oxalate for extraction, Bourne and associates (12)

Table 5. Pectic Substances

Protopectin: water-insoluble, consisting of methyl-
               polygalacturonide chains, parent sub-
               stance which on hydrolysis yields
               pectin or pectinic acids

Pectic Acids: water-soluble polygalacturonic acids
               free of methyl ester groups;
               salts are normal or acid pectates

Pectinic Acids: water-soluble polygalacturonic acids
                 with considerable methyl ester groups

Pectin: water-soluble, greatly esterified (~ 80%)
         polygalacturonic acids

examined the cured leaf and fresh leaf pectic substances in Bright tobacco. The analytical data showed that cured leaf pectic sub-stances were predominantly polygalacturonic acid material, while fresh leaf material had other carbohydrate substances associated with it; thus, curing results in a degradation and loss of the neutral sugars attached to the acidic chain. Hydrolysis products usually contained d-galacturonic acid, d-galactose, l-rhamnose, l-arabinose and smaller quantities of glucose, xylose, and fucose. The isolation of alpha-1,4 linked di-, tri-, and tetra-galacturonic acids is indicative of a 1,4 linkage in the polymer.

A procedure used to obtain the pectic substances of tobacco stems of various types is illustrated in Figure 4 (13). It is interesting to note that 0.1N hydrochloric acid in 2-propanol was used to solubilize the calcium associated with the pectates (Fraction A). The residue was sequentially extracted with boiling water, dilute alkali, and ammonium oxalate to yield fractions with the indicated galacturonic acid (GU) contents, for a total of 12.3% galacturonic acid in Bright stems. The isolation of free pectinic acid (14) from flue-cured tobacco is shown in Figure 5. This was claimed to be fairly pure polygalacturonic acid, as evidenced by the determined equivalent weight of 187, the

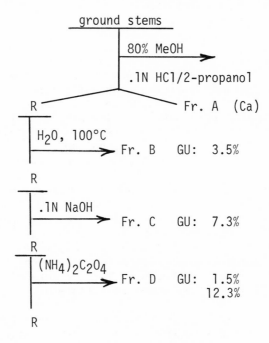

Fig. 4. Extraction of Pectins of Bright Tobacco Stems.

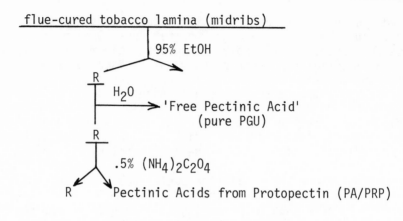

| | yield | | MWx 10^4 | | |
	L	MR	L	MR	EQ W
free PA	2.3	3.2	5.8	5.5	187
PA/PRP	11.2	11.7	4.4	4.0	215-28

Fig. 5. Extraction of Flue-cured Tobacco Lamina (L) or Midribs (MR) for free Pectinic Acids (PA).

theoretical value for anhydrogalacturonic acid being 176. These free pectinic acids were isolated with water from lamina with a yield of 2.3% and a molecular weight of about 58,000. Midribs yielded free pectinic acids at 3.2% with a molecular weight of 55,000. The rest of the pectic substances were obtained in 11 to 12% yield with molecular weights of about 40,000, calculated from intrinsic viscosities. So far it appears that numerous types of pectic substances exist and that the structure of isolated pectic substance depends on the solvent and conditions of extraction.

A related polysaccharide (15) associated with brown pigments has been isolated from aged burley tobacco and has been found to contain the following: glucuronic acid, galactose, arabinose, rhamnose, protein, calcium and magnesium. In light of the preceding discussion, this would appear to be a mixture of calcium and magnesium pectates, loosely bonded to a protein moiety.

The last group of high molecular weight constituents of tobacco that will be discussed are the polymeric polyphenols, which include lignin and the brown pigments. The chemistry of lignin is quite extensive and has been thoroughly studied by chemists of the paper industry. Natural lignin, as it occurs in wood, is called

protolignin. So far, no method has been devised for isolation of
the lignin in a form identical with that in its natural state.
All isolation procedures produce changes in the natural lignin;
and so all isolated lignin preparations are artifacts and, depending upon the severity of the procedure, may or may not resemble
lignin as it occurs in the plant. Many formulations (16) have
been proposed for natural lignin, based on numerous chemical reaction products. It is essentially composed of para-hydroxy, meta-
methoxy phenyl propane units, with the aliphatic side chain in
various states of oxidation, and bonding is between both aliphatic
and aromatic moieties through carbon-carbon or ether linkages. An
extraction procedure has been devised that involves the removal of
most of the tobacco constituents (17). It involves sequential
extraction of the tobacco with 95% ethanol, alcohol-benzene, 1%
hydrochloric acid, 5% sulfuric acid and water. Crude lignin occurs
in tobacco to the extent of about 3.5% and has a methoxyl content
of 12 to 17%.

The more interesting polymeric polyphenols of tobacco are the
brown pigments. These pigments are not present in green leaf but
are formed on curing, and the color of the cured tobacco is due to
them. Most methods of pigment extraction are similar (18,19,20).
The tobacco may be first extracted with organic solvents like
acetone, ether and alcohol, and then treated with an alkaline
phosphate buffer or sodium hydroxide solution to extract the brown
pigments. The extract is then acidified with acetic acid or
hydrochloric acid, which precipitates most of the pigment material.
There are brown pigments which are soluble in water, in buffer
systems, or in strong alkali and, depending upon which solvent is
used, yields from 0.3 to 4% are obtained. Table 6 shows the composition of these pigments. Elemental analysis shows the presence

Table 6.   Pigment Composition

1. Analysis: C, H, N, S, Fe

2. Chlorogenic acid, rutin
    scopolin, scopoletin

3. Protein - amino acids

4. Alkaloids: nicotine, pyridine
    derivatives

5. Aliphatic acids

6. Silicone - methylsiloxane

of C, H, N, S, and Fe. Hydrolytic reactions produce products
indicative of the presence of such polyphenols as chlorogenic
acid, rutin, scopolin, and scopoletin. Usually, only chlorogenic
acid and rutin are found. The presence of a protein in the pigment
structure is inferred from the finding of 18 to 20 amino acids upon
hydrolysis. More drastic degradations, like potassium hydroxide
fusion at high temperatures and pressure, yielded numerous nicotine
and pyridine derivatives, aliphatic acids, and a silicone, identified as a methylsiloxane (21).

After examination of all data, several generalizations can be
made. Numerous observations have led to the conclusion that these
pigments arise from the enzymatic oxidation of plant phenols,
through a process termed the "browning reaction". This is a general
term used to describe the formation of brown substances in all plants
that are undergoing degradation due to injury or senescence. An
enzyme, polyphenoloxidase (22), is involved in this browning
reaction and oxygen is required to oxidize the o-dihydroxy phenolic
entity to an active intermediate, an o-quinone. The rate of browning increases with temperature and involves usually polyphenols like
chlorogenic acid or rutin. From the degradative experiments, it
appears that other leaf constituents like aliphatic acids or alkaloids are also involved. Also, the reaction proceeds slowly in the
leaf, but more rapidly upon cellular disruption, requires molecular
oxygen and its rate increases with a rise in temperature.

Pigments from various tobacco types apparently are similar in
composition and their molecular weights range from less than 2000
to greater than 100,000 (23). Variation in yield of pigments extracted depends upon the solvent used, since greater amounts are
removed with ammonium or sodium hydroxide than with water. This
may indicate that the pigments exist in the leaf as insoluble metal
salts or acids which are converted to the more soluble sodium salts.
The acidic nature of the pigments is indicated by the fact that they
can be precipitated from solution by the addition of acid at a pH
of 4 or less. From the numerous components it is apparent that the
biosynthesis of the pigments proceeds not only through the browning
reaction but through other condensation mechanisms. I believe that
the structure of the pigment is a network of chains of phenolic
constituents, strongly bonded to each other through carbon-carbon
and carbon-oxygen bonds. Other components like the alkaloids are
probably attached in a random manner, but also through strong
carbon-carbon bonds. The protein, however, is loosely bonded
through amide linkages with properly oriented carboxyl groups and
can be easily removed by acid hydrolysis.

Table 7 shows a comparison of properties between the brown
pigments of leaf and similar, but more complex pigments isolated
from cigarette smoke condensate (24). These smoke pigments contain

Table 7. A Comparison of Smoke and Leaf Pigments

	LP	SP
Chlorogenic acid	+	+
Glucose, rhamnose	+	-
Iron, silicon	+	+
Amino acids	+	+
Alkaloids	+	++
Yield	3-4%	5%
Hydrolysis resistance	±	++
Molecular weight, major	30,000	100,000

only chlorogenic acid, no rutin, and have the same constituents as leaf pigment but much more alkaloid content. Apparently, the leaf pigments are mechanically transferred into the smoke as aerosol particles and during this transfer they polymerize and interact with other smoke bases to form the larger, more complex smoke pigments. Whereas, the major molecular weight band of leaf pigment is about 30,000 that of smoke pigment is greater than 100,000.

In conclusion, I feel that there is still much creative thinking and research that must be done to unravel the mysteries and structures of these high molecular weight constituents of tobacco.

References

1. L. C. van Loon and A. van Kammen, Phytochem. 7, 1727 (1968).
2. N. Kawashima and E. Tamaki, Phytochem. 6, 329 (1967).
3. N. Kawashima, H. Fukushima and E. Tamaki, Phytochem, 6, 339 (1967).
4. J. J. Sheen and B. I. Townes, Beitr, Tabakforsch. 5 (6), 285 (1970).
5. N. Kawashima, A. Imai and E. Tamaki, Plant and Cell Physiol., 8, 595 (1967).
6. N. Kawashima and S. G. Wildman, Am. Rev. Plant Physiol., 21, 325 (1970).
7. N. Kawashima and S. B. Wildman, Biochem. and Biophys. Res. Comm., 41 (6), 1463 (1970).
8. M. Samfield and M. G. Christy, Tobacco Science, 4, 38 (1960).

9. H. Jacin and R. J. Moshy, J. Agr. Food Chem., 16, 669 (1968).
10. T. P. Gaines, Tobacco Science, 14, 164 (1970).
11. M. Phillips, F. B. Wilkinson and A. M. Bacot, JAOAC, 36, 1157 (1953).
12. E. J. Bourne, J. B. Pridham and H. G. J. Worth, Phytochem., 6, 423 (1967).
13. H. Jacin, R. J. Moshy and J. V. Fiore, J. Agr. Food Chem., 15 (6), 1057 (1967).
14. R. de la Burde and S. F. Norman, Tobacco Science, 12, 236 (1968).
15. H. E. Wright, Jr., W. W. Burton and R. C. Berry, Jr., Phytochem., 1, 125 (1962).
16. I. A. Pearl, The Chemistry of Lignin, Marcel Dekker, Inc., N. Y., 1967.
17. M. Phillips and A. M. Bacot, JAOAC, 36, 504 (1953).
18. O. T. Chortyk, W. S. Schlotzhauer and R. L. Stedman, Beitr. Tabakforsch., 3, 421 (1966).
19. H. E. Wright, Jr., W. W. Burton and R. C. Berry, Jr., Arch. Biochem. and Biophys., 86, 94 (1960).
20. J. S. Jacobson, Arch. Biochem. and Biophys., 93, 580 (1961).
21. M. Dymicky, O. T. Chortyk and R. L. Stedman, Tobacco Science, 11, 42 (1967).
22. J. A. Weybrew and R. C. Long, Tobacco Science, 14, 167 (1970).
23. O. T. Chortyk, Tobacco Science, 11, 137 (1967).
24. M. Dymicky and R. L. Stedman, Phytochem., 6, 1025 (1967).

PROJECTED CHANGES IN THE COMPOSITION OF BRIGHT

(FLUE-CURED) TOBACCO

J. A. Weybrew, W. G. Woltz, and R. C. Long[1]

North Carolina State University, Raleigh, 27607

The distinguishing characteristics of flue-cured tobacco -- its bright yellow color, its low nicotine and high sugar contents -- are mostly the result of the process by which it is cured but also of cultural management and genetics. However, the art of growing and curing bright tobacco has not changed in any major way since its inception (2); therefore the mass of compositional data that has accumulated (1, 5) would not be atypical of the current crop. These generalizations do not imply that all lots of tobacco are identical. Large year-to-year and location-to-location differences do occur; these are mostly attributable to weather and, specifically, to the total amount and distribution of rainfall during the period of rapid growth (June in North Carolina). Blending balances out these differences.

So, instead of concerning ourselves over the de facto composition of tobacco -- this is history anyway -- it might be more provocative to look into the future; to take cognizance of changes that likely will take place in tobacco production; to discuss the pro's and con's of several alternatives; and finally to predict the cumulative effects of these changes on the composition of cigarettes by the 1980's.

A TIME OF CHANGE

Tobacco production traditionally has been a hand operation. Harvesting, in particular, requires a large number of laborers, but only for about six to eight weeks during the summer. "Barning" tobacco, i.e., filling a conventional barn in one day,

35

requires a crew of 13 people -- four "primers", two "sledders", four "handlers", two "loopers", and one "flunkey". This labor force, once abundant in the South, has essentially vanished. Many tobacco hands have migrated to full-time jobs in the cities.

Thus machines will now have to perform the tasks formerly done by hand. The transition to mechanization is just beginning. But mechanization encompasses more than just a mechanical harvester. The curing structure and more specifically the curing container, but not the curing process, will have to be adapted to the condition of the leaves as they are delivered from the harvester. Certain agronomic practices may be modified to allow the machine to operate most efficiently. It is not inconceivable that the cured product will have lost its identity as a leaf; this would necessitate drastic changes in the marketing procedure, possibly even private on-farm sales. Several aspects of the total problem are being researched separately, not only at North Carolina State University, but at the experiment stations of the other tobacco producing states as well. The evolution of the final new system will be gradual.

Although a number of commercial machines such as the one shown in Figure 1 have been in use for several years, they hardly merit the label of mechanical harvesters. Machines of this type are more aptly categorized as mobile hand-harvesters. As this two-row machine moves through the field, four seated primers pick the leaves and attach them, in groups of three, to clips on the elevator chain. The leaves are lifted to the overhead platform where they are tied onto sticks as in conventional curing. Periodically the filled sticks are transported and hung in the barn. This operation still requires a crew of 13.

Mechanical Harvesters. Developmental research on mechanical harvesters started at North Carolina State University in the early 1950's (6). The heart of any mechanical harvester is the device that breaks the leaves from the stalks. Our agricultural engineers built and tested several "defoliators" (or "pickers", or "strippers") before opting in favor of the one shown diagrammatically in Figure 2. Two counter-rotating rubber augers (the nearer one is not shown) close around the stalk and the blades wipe the leaves from the stalk as the machine moves forward. The whole picker assembly can be raised and lowered hydraulically as a unit to include the lowest leaf. The angle of inclination is also adjustable hydraulically; this determines the height of the swath, viz., the number of leaves that will be harvested. The detached leaves fall onto conveyor belts that elevate them to a platform at the rear where they fall randomly into some suitable receptacle.

Figure 1. The Long Harvester. Machines of this type are con-
           venience devices; they only provide mobility for hand-
           harvesting.

     The machine pictured in Figure 3 is a mechanical harvester
that was designed by engineers of the R. J. Reynolds Tobacco Company, Winston-Salem, N. C. With this one-row harvester, the single operator can move over the field at the rate of about 3/4
acre per hour. Three mechanical harvesters generally similar to
the one shown here were available in limited numbers in 1971:
"The Roanoke" (Reynolds), Harrington Manufacturing Company, Lewiston, N. C.; "The Powell Combine" (NCSU), Powell Manufacturing Company, Bennettsville, S. C.; and "The Hawk Auto Tobacco Picker",
Eagle Machine Company, Ltd., London, Ontario.

     Even though successful harvesting machines are now available
we doubt that the design, particularly at the delivery end, has
been finalized. How these machines are to be used will greatly
influence design modifications. Let's consider some of the alternatives and, to do this, let's contemplate a tobacco plant that is
5-feet tall with 18 leaves uniformly spaced in a spiral array.

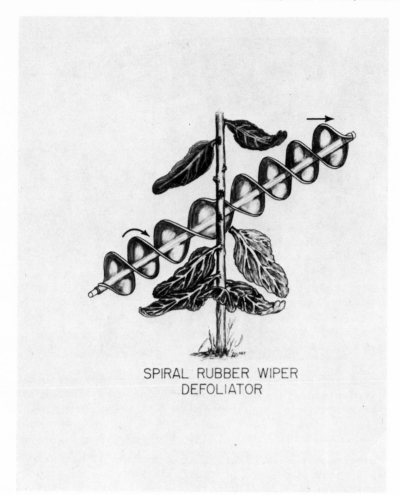

SPIRAL RUBBER WIPER
DEFOLIATOR

Figure 2. Drawing of the tobacco "defoliator" developed by agri-
          cultural engineers at North Carolina State University.
          Two of the commercial mechanical harvesters use this
          picker. From Suggs, (6).

One alternative might be to harvest all leaves in a single pass
("once-over" harvesting), presuming that the pickers now are or
could be made equal to the task. Since all of the leaves will
have been removed, what happens to the plant as the machine passes
over it is of no consequence. Thus, the center of gravity could
be lowered and the machine could be narrowed to straddle only the
one row. This design advantage however is negated by other considerations. Either the conveyor system would have to be designed
to handle the large volume of leaves, or the ground speed of the
machine would have to be reduced to a creep. In addition, the
storage receptacle would have to be excessively large, or the rows
would need to be impractically short; 500 pounds of green leaves
would collect every 150 yards. Worse still, the mixed maturities

Figure 3.   One-row, one-operator mechanical harvester designed and built by engineers of the R. J. Reynolds Tobacco Company.   About 10 of these machines were manufactured by the Harrington Manufacturing Company, Lewiston, N. C. and sold under the trade-name "Roanoke" in 1971.   (Photograph supplied by R. J. Reynolds Tobacco Company)

represented by the leaves collected on once-over harvesting could
not be tolerated; if the bottom leaf was ripe, the uppermost leaf
would be at least a month underripe. Very young leaves cannot be
cured at all; most certainly this mixture of leaves could not be
cured satisfactorily. And even if they could be cured, the final
product would be an inseparable mixture of stalk positions; this
would greatly restrict the manufacturers' latitude for blending.

Having thus rejected once-over harvesting, the remaining alternative is multipass harvesting. If only a few of the lowest leaves
are to be harvested, this automatically imposes the requirements of
high-clearance and sufficient width for stability. Moreover, the
passage of the machine must leave the plant undamaged. How many
leaves should be taken at each pass? The optimum machine efficiency is realized at some inverse compromise between number of leaves
per pass and ground speed. Hand-primed simulation experiments have
demonstrated repeatedly that six-leaf harvests from certain varieties can be cured satisfactorily and sell without penalty. While
obviously six leaves harvested simultaneously cannot all be as
ripe as the conventional three-leaf harvest, still our own unpublished data have shown that a large panel of smokers decisively
preferred cigarettes made of tobaccos harvested one-week underripe
over comparable cigarettes of overripe tobacco. In the multi-har-
vest scheme, successive harvests can be deferred to allow the remaining leaves to ripen further. This plan also retains some seg-
regation by stalk position. The uppermost priming would however
present some problem due to the limberness of the stalk.

Agronomic Adaptations. But the plant need not be five feet
tall. Plant breeders could easily develop low-statured varieties,
but this would usually be expressed through a shortening of the
internodes. With closely-spaced leaves, however, one might predict that one or more by-passed leaves just above the point of
insertion of the picker would be damaged. Shortening can be ef-
fected much more simply, merely by topping the plant lower. Comparative analyses of samples from the authors' management experiments[2] show the effects of topping heights on the yield, value,
and chemistry (total alkaloid, reducing sugars, and starch) of the
cured leaf (Table 1). At constant leaf numbers, e.g., the product
of the leaves/plant x plants/acre, it is not surprising that the
yields were not different ("Conventional" versus "Low-Topped");
neither was it unexpected that yields would be depressed propor-
tionately when one-third of the leaves were discarded in the "Cut-
Back" treatment. The higher alkaloid content of the low-topped
tobacco reflects the higher ratio of root tips (nicotine "facto-
ries") to leaves ("warehouses"). The higher alkaloids of cut-
back tobacco is the result of a degree of over-fertilization of
the fewer remaining leaves. The quality of the low-topped tobacco, as evidenced by its value per hundredweight, was fully equal

Table 1

SOME ASPECTS OF CULTURAL MANAGEMENT: LEAF POPULATION

Variety, C-254   Oxford, 1970

Imposed Variables:

Identification	Conventional	Low-Topping	Cut-Back
Treatment No.	13	1	10
Leaf Density, Lv/Pl x Pl/A	18 x 6336	12 x 9504[a]	12 x 6336[b]
Harvests, No. x Lvs/Hv	6 x 3	3 x 4	1 x 12

Measured Responses:

Yield, Lb/A	2120	2116	1487
Value, $/Cwt	77.60	78.27	77.39
% Total Alkaloid (Wt'd Avg)	4.05	4.33	4.71
% Reducing Sugar (Wt'd Avg)	12.0	12.3	11.2
% Starch (Wt'd Avg)	1.58	1.17	0.91

[a] Topped after the 12th leaf had unfurled from bud.
[b] Broken back to 12 leaves after flowering.

to that of conventionally topped tobacco.

In another treatment, low-topped tobacco was hand-harvested "once-over" to simulate one of the alternatives of mechanical harvesting. The depressed yield (relative to "low-topped") reflects the deterioration of the overripe leaves plus the interrupted growth of the upper leaves (Table 2). The lower value is mostly indicative of the poorer curability of the immature upper leaves.

Axillary branches, "suckers", constitute a potential obstacle to mechanical harvesting of tobacco. Suckers, if present in appreciable numbers and size, would interfere with the operation of the pickers; they would clog the conveyors; they are impossible to cure; and they would contaminate the product. Suckers must be eradicated or utterly suppressed. Attempts to breed a suckerless tobacco have shown little promise. One chemical suppressant has given nearly perfect sucker control; the search for others

Table 2

SOME ASPECTS OF HARVEST MANAGEMENT: ONCE-OVER

Variety, C-254   Oxford, 1970

Imposed Variables:

Identification	Low-Topped	Once-Over
Treatment No.	1	3
Leaf Density, Lv/Pl x Pl/A	12 x 9504[a]	12 x 9504[a]
Harvests, No. x Lvs/Hv	3 x 4	1 x 12

Measured Responses:

Yield, Lb/A	2116	1943
Value, $/Cwt	78.27	75.73
% Total Alkaloid (Wt'd Avg)	4.33	4.17
% Reducing Sugar (Wt'd Avg)	12.3	12.2
% Starch (Wt'd Avg)	1.17	1.10

[a] Topped after the 12th leaf had unfurled from bud.

continues.

Scheduling the Harvest. Up to this point, our attention has focused on how the harvester picks leaves from the plant, and how the plant might be managed to facilitate picking. We need now to consider how to make the best use of the machine over the entire harvesting season. While economics is not a primary intent of this paper, it stands to reason that the major investment in a mechanical harvester ($12 to 15 thousand) can be justified only if the machine operates every day for as long as possible. This will require a close scheduling of the harvest. Once a leaf has ripened, it should be picked within about two days; otherwise deterioration sets in; "holding-ability" is prolonged somewhat by low-topping. A seemingly simple way to schedule ripening would be to stagger the planting. However, a separation in time imposed in May cannot be relied on to persist into July; very often the vagaries of weather will cause two plants set 10-days apart to ripen nearly simultaneously. But staggered harvesting will work more dependably because it is imposed "on the spot". Suppose that it has been decided to harvest six leaves at a time. Harvesting would be started as soon as the 4th leaf had attained minimum ripeness but, for the first two days, only four leaves would be harvested. On the third and fourth days, five leaves would be taken, and on days five and six, the full complement of six leaves would be gathered. Then, on the next go-round two weeks later, leaves 5 through 10 would be ready on the first group of plants, leaves 6 through 11 on the second, and leaves 7 through 12 on the third. Continuing, leaves 11 and 12 on the first group and the 12th leaf on the second would not be scheduled until the 5th week; more practically however, they would probably be included with the second harvest.

Plant breeders can render assistance in scheduling the harvest reliably. Maturation (days-to-flower) is genetically controlled. Current varieties differ by as much as eight days in rate of maturation (4). Earlier or later varieties could be developed on demand.

By planting 36 acres each of two varieties, one 7-days slower maturing, by staggered harvesting six leaves at a time in two passes two weeks apart, and by operating at the rate of 3/4 acre per hour, a mechanical harvester would be in the field for eight hours per day, six days a week, for a month.

This is a tight schedule; one that might easily be knocked out of kilter by a couple of cool rainy days. Some catching up can be accomplished by increasing the swath by one or more leaves. Chemical ripening agents such as "Ethrel"[3], 2-chloroethylphosphonic acid, offers another means for restoring or maintaining the harvesting schedule. Four days after spraying the top of the plant with 180-mg ethrel, all physiologically mature leaves had turned yellow; by one week following treatment, fully ripe leaves had died and fallen off. In addition to the destruction of chlorophyll, treatment with ethrel greatly accelerates respiration. Within three days following treatment, the starch contents of uncured leaves were reduced by 50% relative to untreated plants[4] (Figure 4). These responses seem to be due to ethylene; similar results were induced when an equivalent quantity of ethylene gas was injected into a large plastic bag encasing a plant.

The curing of ethrel-treated leaves, because they had already yellowed on the plant, becomes essentially only a drying process; but drying has to be accomplished speedily though cautiously to prevent browning. This was done successfully; the adjudged value of the ethrel-treated tobacco was not significantly different from the untreated control. The low sugar and starch contents of ethrel-treated tobacco coupled with the 400-lb/A depression in yield bear evidence of the accelerated respiration (Table 3). The apparent

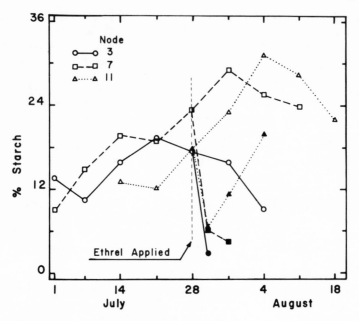

Figure 4.   Starch accumulation in tobacco during growth and matu-
            ration. Application of a ripening agent, Ethrel,
            greatly accelerated respiration. Within three days
            following treatment, 50% of the starch had disappeared.

cencentration of nicotine is a further reflection of excessive
respiratory weight loss.

     Tobacco in south Georgia reacts more favorably to treatments
with ethrel than in North Carolina; more leaves yellow more uniformly from the same dosages. USDA scientist, J. D. Miles, Coastal
Plains Experiment Station, Tifton, Georgia (private communication)
reports only 100-lb depressions in yield from 150-mg treatments.
The reasons for this difference in response are baffling. Possibly it may be variety; Hicks, which because of its disease suscep-
tibility can no longer be grown in North Carolina, is still the
favorite in Georgia. More likely, we think, the difference may be
due to interactions with climate; climatic conditions in Georgia
may allow tobacco to mature more uniformly. On the basis of our
own experiences to date, we would have to regard ethrel as a
standby measure to be used only to overcome emergency delays in
the harvesting schedule.

     Curing Modifications. Recent curing innovations hold the prom-
ise of expediting the conversion to total mechanization. It had

Table 3

SOME ASPECTS OF RIPENING MANAGEMENT: ETHREL

Variety, C-254   Oxford, 1970

Imposed Variables:

Identification	Control	Ethrel-180[a]
Treatment No.	2	4
Leaf Density, Lv/Pl x Pl/A	12 x 9504	12 x 9504
Harvests, No. x Lvs/Hv	1 x 4 plus 1 x 8	1 x 4 plus 1 x 8

Measured Responses:

Yield, Lb/A	2131	1747
Value, $/Cwt	73.55	74.01
% Total Alkaloid (Wt'd Avg)	4.26	4.50
% Reducing Sugar (Wt'd Avg)	15.2	7.4
% Starch (Wt'd Avg)	1.24	0.51

[a]Applied immediately following 1st harvest.

been mentioned in passing that two of the three mechanical harvesters now available deliver the leaves in random disarray. If
the only recourse were to untangle the leaves and tie them onto
sticks for conventional curing, the advantage gained by mechanical
harvesting would be largely voided. But tangled leaves speared
onto tines in bales weighing up to 150 lb. have been successfully
bulk-cured (i.e., with conditioned air under pressure) and sold
for only $0.10 per cwt less than aligned leaves cured similarly
(6). For the past two summers, W. H. Johnson and his associates
in the Department of Biological and Agricultural Engineering,
North Carolina State University, have been experimenting with
cutting green leaves at harvest into sized pieces (3x3 inches) and
then curing these pieces in cages with a forced-air system. For
comparison, whole leaves also in cages and leaves strung onto
sticks were similarly cured. Preliminary results, given in Table
4, are most encouraging; the less-concentrated nicotine in the
sized tobacco suggests some restriction in respiration.

Table 4

SOME ASPECTS OF CURING MANAGEMENT: PRESENTMENT

Oxford, 1970

PRESENTMENT	% Total Alkaloids	% Reducing Sugars	% Total Nitrogen
	Averages of Five Primings		
Whole leaves on sticks	3.09	12.4	2.38
Whole leaves in cages	2.96	14.5	2.44
3x3 in. pieces in cages	2.64	15.0	2.26

The midrib (stem) is an integral part of the leaf; it supports the leaf during growth and conducts water and nutrients into and out of the leaf. Good use was made of this "handle" in traditional tobacco culture. Leaves were primed by the stem and they were tied onto sticks for curing by this handle. It has been estimated that nearly one-half of the curing time and more than one-half of the curing fuel are expended in the "killing out" of the stem. The utility of the stem continued beyond curing. Leaves were formed into bundles for market by the stems, and the integrity of the bundle was retained through redrying, aging, and tipping. In the factory, stems were ultimately stripped out, formerly by hand, but now by threshers. Fifteen years ago, tobacco stems were a waste product of the cigarette factory; they were spread out on lawns and golf-course fairways as fertilizer. Now, however, the separated stems are crushed through rollers, finely cut, and blended back into the cigarette. Their use is justified on the basis that tobacco stems constitute a natural "extender". Evidently the smoker is unaware of their presence or has adjusted to them.

The opportunity is now at hand to dispose of this too-expensive convenience. Only about 15% of the 3x3 pieces contain sections of midrib. These pieces are heavy and their separation seemingly might easily be accomplished on the basis of relative bouyance in air. The fringe of lamina attached to the midrib probably would be sacrificed. It is not difficult to imagine that Dr. Johnson's cutters and air-separators might be incorporated into the conveyor system on the harvester. Discarding the midrib in the field would save the grower curing time and fuel, and the manufacturer could by-pass his threshers.

The curing barn of the future may be nothing more than a rain shelter. The receptacle on the harvester might be a large curing cage, appropriately modified and mounted on wheels. When filled, it might be parked alongside and attached to a plenum supplying conditioned air. It thus becomes the curing unit.

CIGARETTES, NOW AND THEN

Finally, I propose to compare the composition of cigarettes before and after mechanization. Obviously, since no tobacco has been grown and cured under a system incorporating all of the ideas discussed in this paper, this prediction will involve numerous assumptions and approximations.

The constituent tobaccos in an "Average American Blended Cigarette", as determined by tediously picking apart cigarettes of all popular brands, are given in Table 5. The proportions of burley, Maryland, and Turkish tobaccos will be held constant in both hypothetical cigarettes, and their contributions to the chemistry of the blend will be computed from their compositions as reported by Harlan and Moseley (3). It will be assumed that the homogenized tobacco component and the processed stems are also flue-cured.

Table 5

"AVERAGE" AMERICAN BLENDED CIGARETTE

COMPONENT	Abundance (%)
Flue-Cured	43.5
Burley	23.5
Turkish	16.8
Homogenized Leaf	11.0
Processed Stems	3.1
Maryland	2.1

For the "Now" cigarette, the flue-cured component will assume conventional culture, viz., 6336 plants per acre topped at 18 leaves, harvested in 6 x 3 primings, and conventionally cured; its contribution to the blend will be computed from the weighted averages given in Table 1.

The flue-cured component of the "Then" cigarette will assume 9504 low-topped (14 leaves) plants per acre, leaves 3 through 14 to be machine-harvested in 2 x 6-leaf swaths, cut into 3x3 pieces (stem discarded), and cage-cured. The bottom two leaves will be abandoned; they are usually quite inferior and are generously coated with splashed sand. And besides, it would be foolhardy indeed even to consider operating the intricate and expensive picker heads on or in the surface of the soil.

The synthesized compositions of the "Now" and "Then" cigarettes are compared in Table 6. These cigarettes are amazingly alike. In fact, differences larger than these often occur between tobacco fields only miles apart in any year, and are easily balanced out in the blending room. One thing is certain; the taste of the "Then" cigarette will not have suffered from the exclusion of the stem!

Table 6

CIGARETTES, "NOW" & "THEN"

BRAND	% Total Alkaloids	% Reducing Sugars
Now	3.11	8.9
Then	3.16	11.3

FOOTNOTES

[1/] Respectively, Professor of Crop Science, Professor of Soil Science, and Assistant Professor of Crop Science.

[2/] Unpublished. An Annual Report of Accomplishments in Tobacco Research in North Carolina. North Carolina State University, 1967. pp. 155-157.

[3/] Amchem Products, Incorporated, Ambler, Pennsylvania. Mention of a proprietary product is not to be construed as an endorsement to the exclusion of other similar materials.

[4/] Unpublished. An Annual Report of Accomplishments in Tobacco Research in North Carolina. North Carolina State University, 1970. pp. 252-275.

LITERATURE CITED

1. Bacot, A. M. The chemical composition of representative grades of the 1952 and 1954 crops of flue-cured tobacco, including chemical methods. United States Department of Agriculture, Tech. Bul. No. 1225. 1960.

2. Brooks, J. E. The Mighty Leaf. Alvin Redman Ltd., London, 1953. p. 236.

3. Harlan, W. R., and J. M. Moseley. Tobacco. Ency. Chem. Tech., 14:242-261, 1955.

4. Rice, J. C., D. T. Gooden, and E. L. Price. Measured crop performance: Tobacco. North Carolina State University, Res. Report No. 36. 1970.

5. Stedman, R. L. The chemical composition of tobacco and tobacco smoke. Chem. Rev., 68:153-207, 1968.

6. Suggs, C. W. Industry ripe for mechanical harvester. Tobacco, 173(1):17-23, 1971.

EFFECT OF TOBACCO CHARACTERISTICS ON CIGARETTE SMOKE COMPOSITION

T. C. Tso and G. B. Gori

Plant Science Research Division, ARS, USDA, Beltsville,

Maryland 20705; and National Cancer Institute, HEW,

Bethesda, Maryland 20014.

Two groups of experimental cigarettes were used to examine the effect of leaf tobacco characteristics on cigarette smoke composition. Group I consists of 32 straight flue-cured cigarettes, and Group II consists of 23 blend cigarettes involving many variables. Within each group, simple correlations among leaf and smoke variables, and multiple regression of total TPM, tar, smoke nicotine with selected leaf variables are measured. Results appear to suggest that total N, K, and cellulose content in flue-cured type cigarettes are major factors affecting total TPM delivery, while waxes, total polyphenols, and free histidine are major factors affecting total TPM delivery of blend type cigarettes.

Since the publication of the 1964 Surgeon General's Report on Smoking and Health (1), government agencies such as the Agricultural Research Service, USDA, and the National Cancer Institute, HEW, have devoted considerable effort to health-related tobacco research. The major problems are how to relate leaf characteristics to smoke composition, smoke composition to biological activity in animal tests, and animal tests to human smoking. Once these relationships are established, scientists will be able to produce good quality leaf tobacco which can be used for manufacture of cigarettes with desirable smoke products.

51

This research is a joint undertaking which involves many
Federal, State, industrial and independent institutes. The
present report describes some of the data from the first stage of
these investigations - the simple relationships and multiple
regressions among leaf characteristics and smoke composition.
These results were generated from specially produced tobaccos,
from regular or specially blended experimental cigarettes, and from
cigarettes made of reconstituted tobacco.

Materials and Methods

Two groups of samples were used in this study. The first
group (Group I) of 32 samples involves flue-cured type tobacco of
four widely different alkaloid levels and eight stalk positions.
Alkaloid differences in these samples are almost 15-fold,
including 3.55% for SC-58, 2.74% for NC 95, 1.24% for Coker 139,
and 0.24% for LN-38. Each stalk position represents a composite
sample of 2-4 leaves from 800 plants. These tobacco samples were
produced by Dr. James F. Chaplin at Kinston, North Carolina. Leaf
analysis was made in various collaborating laboratories, and each
group of data will be published separately by individual scientists who conducted the study. Tobacco leaf from each of these 32
samples was made into cigarettes. The cigarette smoke studies
were conducted by the American Health Foundation, New York. The
second group (Group II) of samples involves 23 experimental
cigarettes. These samples are a part of the study by the Tobacco
Working Group, Lung Cancer Task Force, NCI, HEW. These samples
include regular and specially blended leaves, stems, or various
formulations of reconstituted tobacco sheets. Smoke studies of
these cigarettes were conducted by the Oak Ridge National Laboratory,
Oak Ridge, Tennessee.

Determinations were made, as follows: Leaf physical properties (Group I only):   fire-holding capacity, color, leaf
thickness, trichomes, moisture equilibrium and specific volume.
Leaf chemical properties: sugars, starch, lignin, cellulose, ash,
α-amino N, phenolic compounds, nitrates, organic acids, amino
acids, alkaloids, nitrogenous compounds, inorganics, phytosterols,
fatty acids, lipids, waxes, pesticide residues, and others.
Cigarette physical parameters: weight, puff number, burn rate.
Smoke delivery: smoke solids, water, nicotine. Cigarette smoke
chemistry: acetaldehyde, acrolein, phenol, nicotine, HCN,
benzo(a)pyrene (BaP) and others.

Data generated from Groups I and II were computerized
separately. Within each group, all data were used to calculate
simple correlations among all combinations of variables. The
multiple regression of a selected smoke constituent, such as total

particulate matter (TPM) or nicotine, or acrolein, with a selected set of variables from leaf studies such as leaf thickness, alkaloids, phytosterols, polyphenols, cellulose, starch, wax, inorganics, or others, was calculated.

Results and Discussion

Group I - flue-cured tobacco samples and cigarettes

Part of the results from Group I were discussed previously (2) with emphasis on varietal and stalk position effects on certain physical and chemical leaf properties. Detailed data will be published separately. Briefly, there are significant differences among varieties in certain physical and chemical characteristics, including leaf thickness, alkaloids, total volatile bases(TVB), petroleum ether extracts, fatty acids, phytosterols and oxalate. There are also many significant variations in physical and chemical properties among leaf positions. For example, there is a consistent increase in leaves from bottom to top stalk positions of number of trichomes, leaf thickness, alkaloids, total nitrogen, total volatile bases. A similar but not very consistent trend was observed in oxalate and polyphenol contents. The total fatty acid content, petroleum ether extract, and phytosterol content, however, are higher in leaves from the middle stalk positions.

Many simple correlations are above 0.7 and are significant at the 1% level. Some of them are of particular interest. For example, there was a negative correlation between cellulose and scopoletin, between cellulose and sugar, between ash and polyphenols, between K and scopoletin, between Mg and number of trichomes, between static burning rate and scopoletin. A positive correlation was observed between K and stigmasterol, fire-holding capacity and number of trichomes, number of puff and leaf thickness, tryptophan and total free fatty acids. The dry TPM is positively correlated with, among many other things, scopoletin, number of trichomes, and oxalate, but negatively correlated with leaf pH, stigmasterol, and fire-holding capacity.

Based on these and other correlation data, multiple regressions of dry TPM of sample Group I with several leaf variables including polyphenols, phytosterols, pH value, total nitrogen (N), potassium (K), and cellulose in each variety were determined, as shown in Table 1 (Equation I). An increase of six more leaf variables, as shown in Table 2 (Equation II) merely increased the R^2 value 0.2%, indicating that sugar, trichome, leaf thickness, lipid residue, oxalate and malate contents do not provide additional information to those six variables already included in Equation I. The predicted dry TPM values from leaves of each

Table 1 - Estimation of Dry TPM of Sample Group I
(Prediction Equation I)

Est. Dry TPM (mg/cig.) = -117.79134 (constant)
                    + 16.81397 X Total polyphenols (%)
                    - 21.52252 X Total phytosterols (mg/g)
                    + 73.11366 X pH value
                    +103.82507 X Total nitrogen (%)
                    - 60.28903 X Potassium (%)
                    - 2.27422 X Cellulose (%)

Coef. of determination R^2 X 100 = 96.8%

Table 2 - Estimation of Dry TPM of Sample Group I
(Prediction Equation II)

Est. Dry TPM (mg/cig.) = 170.16504 (constant)
                    + 17.62296 X Total polyphenols (%)
                    - 17.00079 X Total phytosterols (mg/g)
                    + 50.87225 X pH value
                    - 5.22681 X Sugar (%)
                    - 0.09503 X Trichome (within 3 mm diameter)
                    + 16.50523 X Leaf thickness (mm)
                    + 92.31470 X Total nitrogen (%)
                    - 55.31998 X Potassium (%)
                    - 8.09804 X Lipid residue (%)
                    - 0.18060 X Oxalate (Meq/g)
                    - 81.77171 X Malate (Meq/g)
                    - 2.64672 X Cellulose (%)

Coef. of determination R^2 X 100 = 97.0%

stalk position within the variety NC-95, in comparison with
actually determined values, are shown in Table 3. These data appear
to be quite comparable except one, the second leaf position, which
has a 10.58% deviation. Based on stepwise regression studies, the
dry TPM delivery of Group I samples appears to be very dependent
on total N, K, and cellulose content in the leaf. Polyphenols,
phytosterols, and leaf pH may affect the TPM level, but are not as
decisive as N, K, or cellulose.

Table 3 - Comparison of Actual and Predicted Dry TPM values in
Samples From Group I
(Var. NC-95 based on Equation II)

Leaf Position	Actual (A)	Predicted (B)	Deviation	
			(A-B)	% of A
1	17.18	18.06	-0.88	-5.12
2	23.33	20.86	+2.47	+10.58
3	33.00	31.28	+1.72	+5.21
4	34.02	34.29	-0.27	-0.79
5	38.85	39.55	-0.70	-0.18
6	41.70	43.59	-1.89	-4.53
7	45.58	45.65	-0.07	-0.01
8	48.94	49.53	-0.59	-0.12

Group II - blended regular and reconstituted cigarettes

The 23 experimental cigarette samples varied in blending
formulation, paper porosity, additive cuts and also differences
in methods of sheet reconstitution. It is not the purpose of this
report to examine the smoke differences as results of such formulation, but rather to study the smoke characteristics in relation to
changes in general leaf composition. Data from all 23 samples
are used to calculate simple correlation and multiple regressions.

Simple correlations which are above 0.7 and at 1% significant
level occurred in many cases. Among the more interesting ones are
positive correlation between 3β-hydroxysterols and TPM, smoke
nicotine, tar, and leaf wax; between total polyphenols and TPM,
petroleum ether extracts, and 3β-hydroxysterol; between wax and
TPM, smoke nicotine, and tar; between leaf alkaloid level and wax,
and petroleum ether extracts. Significant negative simple
correlations were found between acetaldehyde or acrolein and 3β-
hydroxysterols, wax, and total alkaloids. Positive correlation
between smoke TPM and wax, petroleum ether extracts and total
alkaloids are also very significant. Detailed correlation results
are shown in Table 4 (3β-hydroxysterol); Table 5 (polyphenols);
Table 6 (wax); Table 7 (total alkaloids); Table 8 (total nitrogen);
and Table 9 (smoke TPM).

Table 4 – Correlation Between 3β–hydroxysterol (mg/g) and
    other Variables from Sample Group II

        Puffs to reach 23 mm butt length   .770
        TPM (mg/cig.)   .751
        Nic/Cig. (mg/cig.)   .831
        Tar (mg/cig.)   .753
        Wax (% tob. wt.)   .897
        Pet. ether extracts   .950
        Ratio Nic/TVB   .739
        Acetaldehyde (μg/liter smoke)   −.718
        Acrolein (μg/liter smoke)   −.713

Puffs to reach 23 mm butt length	.770
TPM (mg/cig.)	.751
Nic/Cig. (mg/cig.)	.831
Tar (mg/cig.)	.753
Wax (% tob. wt.)	.897
Pet. ether extracts	.950
Ratio Nic/TVB	.739
Acetaldehyde (μg/liter smoke)	−.718
Acrolein (μg/liter smoke)	−.713

Table 5 – Correlation Between Total Polyphenols(%) and
    other Variables from Sample Group II

Puffs to reach 23 mm butt length	.740
TPM (mg/cig.)	.703
Ratio Nic/TVB	.854
Pet. ether extracts	.750
3β–Hydroxysterol (mg/g)	.781

Table 6 – Correlation Between Wax (%) and other Variables
    from Sample Group II

Puffs to reach 23 mm butt length	.774
TPM (mg/cig.)	.792
Nic/Cig. (mg/cig.)	.897
Tar (mc/cig.)	.780
Acetaldehyde (μg/liter smoke)	−.825
Acrolein (μg/g tob. burned)	−.756

Table 7 – Correlation Between Total Alkaloids (%) and
    other Variables from Sample Group II

Acetaldehyde (μg/liter smoke)	−.802
Acrolein (μg/liter smoke)	−.724
Moisture equilibrium	−.719
Wax (% tob. wt.)	.863
Petroleum ether extracts	.952

Table 8 – Correlation Between Total Nitrogen and other
Variables from Sample Group II

Acrolein (µg/cig.)	–.763
Wax (% tob. wt.)	.755
Pet. ether extracts (% tob. wt.)	.759
% α-Amino N	.758
Cholesterol (mg/g)	.741

Table 9 – Correlation Between Smoke TPM (mg/g tob. burned)
and other Variables from Sample Group II

Wax (% tob. wt.)	.825
Petroleum ether extracts (% tob. wt.)	.739
Total alkaloids (%)	.727

Three multiple regressions were calculated with the Group II
samples. Table 10 shows multiple regression of TPM with 10 leaf
variables including total alkaloid, phytosterols, polyphenols,
total nitrogen, cellulose, free phenylalanine, histidine, threonine,
oxalate, and wax. Stepwise regression examination showed that wax
is the most important variable; polyphenols and histidine also
were important in contributing to TPM levels as shown in Table 11.

Table 10 – Estimation of TPM of Sample Group II

Est. TPM (mg/cig.) = 71.32877 – 7.33910 X Total alkaloids (%)
+ 4.81838 X Total phytosterols (mg/g)
+ 1.71153 X Total polyphenols (%)
–20.73401 X Total nitrogen (%)
– 0.50334 X Cellulose (%)
–11.37499 X Phenylalanine (µM/g)
+19.14728 X Histidine (µM/g)
+ 1.18948 X Threonine (µM/g)
+27.14696 X Oxalate (%)
+62.33328 X Wax (%)

Coef. of determination R^2 X 100 = 75.3 %

Table 11 – Stepwise Regression Phase of TPM Estimation
Sample Group II

Step I Est. TPM (mg/cig.) = 13.71721
+ 104.73796 X Wax (%)

Coef. of determination R^2 X 100 = 62.7%

Step II Est. TPM (mg/cig.) = 10.46640
+ 2.17632 X Total polyphenols (%)
+ 80.83185 X Wax (%)

Coef. of determination R^2 X 100 = 65.8%

Step III Est. TPM (mg/cig.) = 6.33843
+ 3.24025 X Total polyphenols (%)
+ 9.59483 X Histidine (μM/g)
+ 48.36261 X Wax (%)

Coef. of determination R^2 X 100 = 68.7%

A very similar regression was observed with smoke tar and the
same set of leaf variables. Tables 12 and 13 show, respectively,
the multiple regression and stepwise regression involving these
variables.

Table 12 – Estimation of Smoke Tar of Sample Group II

Est. "Tar" (mg/cig.) =   57.44458 (Constant)
- 5.99162 X Total alkaloids
+ 3.95704 X Total phytosterols
+ 1.68472 X Total polyphenols
-16.82582 X Total nitrogen
- 0.42282 X Cellulose
- 8.15391 X Phenylalanine
+16.31693 X Histidine
+ 0.83101 X Threonine
+28.98094 X Oxalate
+44.25793 X Wax

Coef. of determination R^2 X 100 = 74.7 %

Table 13 - Stepwise Regression Phase of Smoke Tar Estimation,
Sample Group II

Step I Est. Smoke Tar (mg/cig.) = 11.41901
+ 85.19737 X Wax

Coef. of determination R^2 X 100 = 60.9%

Step II Est. Smoke Tar (mg/cig.) = 8.26095
+ 2.11423 X Total polyphenols
+ 61.97325 X Wax

Coef. of determination R^2 X 100 = 65.2%

Step III Est. Smoke Tar (mg/cig.) = 4.96706
+ 2.96319 X Total polyphenols
+ 7.65613 X Histidine
+ 36.06462 X Wax

Coef. of determination R^2 X 100 = 68.0 %

The multiple regression of smoke nicotine content with the
same set of leaf variables was determined, as shown in Tables 14
and 15. The most important variable is again leaf wax content;
histidine and leaf alkaloid content are of next importance.

Table 14 - Estimation of Smoke Nicotine of Sample Group II

Est. Nicotine (mg/cig.) = 0.54456 (Constant)
-0.37629 X Total alkaloids
+0.97428 X Total phytosterols
+0.02867 X Total polyphenols
-0.77667 X Total nitrogen
+0.01127 X Cellulose
-0.83908 X Phenylalanine
+0.72764 X Histidine
+0.11741 X Threonine
-1.38410 X Oxalate
+5.51457 X Wax

Coef. of determination R^2 X 100 = 89.9%

Table 15 – Stepwise Regression Phase of Smoke Nicotine, Sample
Group II

Step I Est. Nicotine (mg/cig.) = 0.10979
+ 10.28943 X Wax (%)

Coef. of determination R^2 X 100 = 80.5%

Step II Est. Nicotine (mg/cig.) = -0.12579
+ 0.89032 X Histidine
+ 8.36101 X Wax

Coef. of determination R^2 X 100 = 84.7%

Step III Est. Nicotine (mg/cig.) = -0.39998
+ 0.38201 X Total Alkaloids
+ 0.63744 X Histidine
+ 6.12539 X Wax

Coef. of determination R^2 X 100 = 86.4%

Acetaldehyde and acrolein were found to be positively
correlated with chlorogenic acid and starch. Multiple regression
among these variables was also determined, as shown in Tables 16
and 17.

Table 16 – Estimation of Acrolein of Sample Group II

Est. Acrolein (μg/cig.) = 26.99522
+ 15.42181 X Chlorogenic acid (%)
+ 3.83689 X Starch (%)

Coef. of determination R^2 X 100 = 52.6%

Table 17 – Estimation of Acetaldehyde of Sample Group II

Est. Acetaldehyde (μg/cig.) = 663.24194
+ 311.77148 X Chlorogenic Acid (%)
+ 68.97284 X Starch (%)

Coef. of determination R^2 X 100 = 56.1%

Comparative evaluation of Groups I and II

It is of interest to compare the different effects of leaf
tobacco characteristics to smoke composition among these two groups
of samples. As mentioned previously, Group I represents straight
flue-cured cigarettes and Group II consists of blend cigarettes
with many variables.

In leaf tobacco, it is known that various nitrogenous fractions, including total N, total volatile bases, nicotine, and other nitrogenous components generally follow similar trends in their respective levels and are positively correlated to one another. This nitrogenous fraction is usually negatively correlated to leaf pH value, total ash content, especially K. Such a relationship holds true in the leaf analysis of Group I samples. In addition, the TPM delivery of Group I cigarettes is positively associated with total N, free amino acids, sugars, fatty acids, oxalate and polyphenols, but negatively associated with K, fire-holding capacity, stigmasterol, total phytosterol, malate, citrate and cellulose.

On Group II samples, the general effects of total N, total volatile bases (TVB), nicotine, oxalate, K and leaf pH value on TPM delivery are similar to that in Group I, but the intensity of their effects are in a much lesser degree. On the other hand, the positive effects of total polyphenol and free amino acid toward TPM delivery are much increased.

The most significant differences among these two sample groups regarding TPM delivery are from phytosterols, petroleum ether extracts, and especially waxes. Each of these three groups of components is either negatively associated or has no association with TPM delivery from cigarettes of Group I, but are all significantly and positively correlated to TPM delivery from cigarettes of Group II. Tables 18, 19, and 20 list comparative results.

Table 18

COMPARISON OF RESULTS BETWEEN GROUPS I AND II: (A) Similar correlations, but higher in Group I than in Group II

Correlations between dry TPM and:	I	II
Total N	.933**	.537**
Total volatile bases	.812**	.574**
Nicotine	.775**	.727**
Oxalate	.705**	.614**
K	-.889**	-.524**
pH	-.735**	-.264

** 1% significance

Table 19
COMPARISON OF RESULTS BETWEEN GROUPS I AND II: (B)   Similar
correlations, but higher in Group II than in Group I

Correlations between dry TPM and:	I	II
Total polyphenol	.522**	.670**
Phenylalanine	.497**	.559**
Threonine	.220	.621**
Cellulose	-.589**	-.641**

** 1% significance

Table 20
COMPARISON OF RESULTS BETWEEN GROUPS I AND II: (C)   Reverse
correlations in Groups I and II

Correlations between dry TPM and:	I	II
Total phytosterol	-.499**	.742**
Stigmasterol	-.714**	.718**
Petroleum extracts	-.033	.739**
Wax	-.299	.825**
Histidine	-.023	.563**

** 1% significance

It is known that straight flue-cured cigarettes and blend
cigarettes differ in smoke acidity because of differences in main
constituents of the leaf. This difference may partially explain
the variation in smoke delivery and smoke composition. In addition,
the Group II cigarettes involved many other variables, including
cigarette paper porosity, leaf cuts, blend formula, additives, and
methods of sheet reconstitution. Physical characteristics as well
as chemical properties may all contribute to the widely different
results.

Additional data of smoke analysis is becoming available and
may yield further information in this research.

Acknowledgment

The authors wish to acknowledge the cooperation of many scientists involved in this project, from Federal, State, industrial, and independent institutions. We are especially indebted to E. James Koch, Agricultural Research Service, USDA, for statistical evaluations.

References

1. Smoking and Health Report to the Advisory Committee of the Surgeon General of the Public Health Service, U. S. Dept. of Health, Education and Welfare, PHS #1103, 1964.

2. Tso, T. C., Chaplin, J. F. and Rathkamp, G. Leaf characteristics, smoke composition, and biological activity. I. Characteristics of leaf and smoke from eight stalk positions of four flue-cured varieties. CMA Research Seminar, Philadelphia, Pa. 1971.

Pyrolysis of Tobacco Extracts

W. S. Schlotzhauer[1], E. Barr Higman[1] and I. Schmeltz[2]

[1]Richard B. Russell Agricultural Research Center, ARS,
USDA, Athens, Ga. 30604 and [2]Eastern Marketing and
Nutrition Research Division, ARS, USDA, Philadelphia,
Pa. 19118.

In discussing the role of pyrolysis in tobacco research, we
shall attempt to review briefly the past history of pyrolysis
experiments in smoking and health related studies, and to present
in some greater detail data obtained from several representative
pyrolyses of tobacco leaf extracts.

The last comprehensive review of the chemical composition of
tobacco leaf and tobacco smoke lists more than 1200 individual
components (1). Many of the smoke constituents are either absent
in cured tobacco leaf, or present in quantities too minute to
account for their presence in smoke. Obviously, these constituents
are produced during the smoking process through various thermal
alterations of precursor components in the cured tobacco leaf.
These alterations might be of several varieties; the thermal rup-
ture of a leaf component into smaller fragments is termed
"pyrolysis", while subsequent recombination of these fragments to
form new smoke components is termed "pyrosynthesis". In addition,
leaf components which are transferred into the smokestream essentially unaltered may be considered "distillation" products. In
the present discussion, we are concerned with the formation of new
smoke constituents from cured tobacco leaf, and hence with both
"pyrolysis" and "pyrosynthesis," terms which will be used inter-
changeably.

Many smoke constituents of biological interest, for example,
the tumor-initiator benzo(a)pyrene and other polynuclear aromatic
hydrocarbons (PAH) (2), and the cilia-movement inhibiting volatile
phenols (3), were early recognized as arising in smoke via pyrolytic
reactions. Thus, many researchers sought to identify the leaf
precursors of these particular smoke constituents, and to determine

the pyrolytic mechanisms involved in their production. Some of
the earlier studies were tacitly based on the assumption that one,
or, at most, a few leaf components would serve as precursor for a
specific smoke constituent, for example, benzo(a)pyrene. This
approach was, of course, a gross oversimplification of complex
processes; however, eventually some important precursor-product
relationships were recognized. In this discussion, we should like
to emphasize two of these relationships; first, the hexane-soluble
components of tobacco leaf as pyrolytic sources of PAH (4, 5) and
second, the brown leaf pigments (6) and carbohydrates (7) as
sources of the volatile phenols. These two classes of smoke
constituent are of particular relevance to the question of cigarette
smoking and health.

In discussing the methodology of pyrolysis experiments, we
shall begin with a brief description of the various parameters
selected, and discuss the rationale for these selections. Most
pyrolyses in tobacco research have been conducted in a chemically
inert, preferably nitrogen, atmosphere. Critics of nitrogen-
atmosphere pyrolysis contend that such experiments portray a false
picture of the smoking process by eliminating various oxidative
reactions; however, Newsome and Keith (8) demonstrated that the
cigarette burning cone and immediate vicinity (as evidenced by
the quantities of hydrogen, carbon monoxide, and methane present)
are essentially in a reducing atmosphere. Combustion, naturally,
does play a role in the smoking process; however, if the role of
combustion were predominant, the major products of burning tobacco
(excluding the specific alkaloids) would be carbon dioxide and
water. Actually, the smoke produced can be considered the product
of a very incomplete combustion.

Selection of experimental temperature is, as expected,
critical in any pyrolytic study. Although the reported temperature range within a burning cigarette is wide, considering
an ignition point of 300-400°C and maximum burning temperature in
excess of 900°C, the majority of published pyrolysis experiments
in tobacco chemistry have been performed at temperatures approaching
the upper limits of this range. Touey and Mumpower (9), using
thin precision thermocouple wires, determined peak burning zone
temperatures to be in the vicinity of 860°C, and reported a sharp
thermal gradient behind the cone, smoldering by convection.
Another rationale for performing pyrolyses at relatively high
temperatures was provided by observations that the carcinogenicity
of tobacco pyrolysates apparently increased with experimental
temperature. Wynder et al. (4) demonstrated that tobacco pyrolysates obtained at 880 and 800°C yielded tars of significantly
higher carcinogenicity than those produced at 720 and 640°C.
Pyrolysates obtained at 560°C failed to show activity. As further
evidence of the utility of high temperature pyrolyses in smoking

and health research, the work of J. Lam (10) might be cited.
Lam demonstrated that tobacco paraffins pyrolyzed at 850°C produced
3 to 10 times the yield of various PAH than that produced when the
experiment was performed at 700°C, at 600°C no evidence for the
formation of these compounds was obtained. In the specific experiments on pyrolysis of tobacco extracts described later in this
paper in some detail, the pyrolysis temperature was 860°C.

We will discuss briefly the various apparatus utilized in
pyrolysis experiments. One arrangement (6) consists of a horizontal
vycor or quartz tube positioned in a tube furnace with accurate
temperature control. Samples are inserted within the hot-zone of
the tube in porcelain boats, and the resultant products flushed by
a continual stream of nitrogen into a series of cold traps or
other appropriate collection devices. This is a very fundamental
arrangement; however, more elaborate systems have been described
in the literature, for example, a vertical tube with sample introduced at the top and falling through the hot-zone at a controlled
rate (11), or a movable furnace mechanically carried across the
sample at a set speed (12). Although the researchers choice of
apparatus design has varied, in each case, the principle is
essentially identical. The resulting pyrolysates are generally
less complex than cigarette smoke condensate, and are thereby
fractionated into individual components under less drastic procedures than those required for condensate assay. Fractionation of
tobacco pyrolysates, through various solvent partitionings and pH
manipulations, yields ethereal solutions of neutrals, phenols,
bases, and carboxylic acids suitable for gas-liquid, thin-layer,
or column chromatography and various methods of spectral analysis
including mass, infrared, and ultraviolet-absorption spectroscopy.

Leaving the general discussion of pyrolysis, we shall consider
the specific application of this technique to smoking and health
research and describe several representative studies. Ever since
a British research group, headed by A. J. Lindsey (13), first
confirmed the presence of PAH in cigarette smoke, considerable use
has been made of pyrolytic methods to identify the leaf precursors
of this class of smoke constituent. Table I lists a number of
these investigations.

Paraffins were early suspected precursors of PAH. Lam demonstrated the production of at least 30 such compounds by pyrolysis
of tobacco paraffins at 850°C. Gilbert and Lindsey, and Wynder and
Hoffmann, found further evidence o f the potency of paraffins as
PAH precursors. G. M. Badger used pyrolysis data from individual
paraffins, such as dotriacontane, to propose various free radical
mechanisms in the pathway from aliphatic leaf components to smoke
PAH. Not suprisingly, various researchers verified that the
tobacco phytosterols -- including stigmasterol and β & γ-
sitosterols, which contain an internal phenanthrene skeleton --

Table I. Precursors of Polynuclear Aromatic Hydrocarbons

Precursor	Reference(s)
Paraffins	Lam (10)
	Gilbert & Lindsey (2)
	Wynder & Hoffmann (14)
Dotriacontane	Badger et al. (15)
	Schlotzhauer & Schmeltz (16)
Phytosterols	Wynder & Hoffmann (14)
Stigmasterol	Badger et al. (15)
β-Sitosterol	Schlotzhauer & Schmeltz (16)
Phytol	Schlotzhauer & Schmeltz (16)
Isoprene	Gil-Av & Shabati (17)
	Oro et al. (18)

produce good yields of PAH on pyrolysis. Schlotzhauer and Schmeltz obtained significant amounts of PAH by pyrolysis of the C_{20} isoprenoid alcohol, phytol -- a tobacco leaf constituent. That the isoprenoids of tobacco leaf are important contributors of PAH in smoke is strengthened by the data of Gil-Av and Shabati, and Oro et al., that pyrolysis of isoprene alone produces large numbers of PAH; the latter group identified 64 aromatic hydrocarbons, including at least 19 PAH, in pyrolysates obtained from isoprene in hydrogen atmosphere. Experiments performed by the Agricultural Research Service, United States Department of Agriculture, sought to determine the relative effectiveness of various individual compounds toward pyrolytic production of PAH. The results for several of these test compounds are tabulated in Table II.

It is apparent from these data (Table II), that the structural characteristics of the precursor have a marked influence on the yield of PAH, for example, isoprenoid compounds being significantly

Table II. Yields of Aromatic Hydrocarbons, including PAH from Various Tobacco Leaf Constituents (16)

Compound Pyrolyzed	Structural Features	Relative Yields
Squalene	isoprenoid	2.72
Linolenic acid	unsat. fatty acid	2.21
β-Sitosterol	sterol; isoprenoid	1.62
Phytol	isoprenoid	1.55
Stearic acid	sat. fatty acid	1.25
Dotriacontane	C_{32} aliphatic	1.14
Hexane	C_6 aliphatic	1.00

more effective precursors than simple aliphatic compounds tested.

Researchers headed by R. L. Stedman (19) have conducted extensive studies of the hexane-soluble portion of flue-cured tobacco, and characterized this leaf fraction (approximately 6% dry leaf weight) as consisting of aliphatic and cyclic paraffins, higher fatty acids and their esters, sterols and their esters, terpenes, phthalates, glycerides, and neutral "resins". As a variety of leaf components identified as PAH precursors were seen to be concentrated in this relatively small portion of tobacco leaf, a number of pyrolytic investigations were stimulated with varying results. Wynder et al. (4) demonstrated the carcinogenicity of pyrolysates of the hexane extract; however, Rayburn et al. (20), in seemingly contradictory findings, reported no reduction in PAH levels (measured as absorbance at 385 mμ) in the smoke of hexane-extracted cigarettes. In an effort to clarify this issue, we extracted flue-cured tobacco with hexane and pyrolyzed the unextracted tobacco, the hexane extract, and the tobacco with hexane-solubles removed, each under identical conditions (860°C, nitrogen atmosphere). Benzo(a)pyrene was isolated from the neutral fractions of the three pyrolysates by thin-layer chromatography and quantitatively estimated by absorbance (363,383 mμ) in cyclohexane. The results of this study are presented in Table III.

As the data in Table III indicate, the hexane extract of flue-cured tobacco contributes disproportionately to the total benzo(a)pyrene content of the tobacco pyrolysate; expressed as yield per gram material pyrolyzed, one gram of hexane extract produced 2400 μg of benzo(a)pyrene, whereas the tobacco with hexane-solubles removed produced less than 90 μg per gram pyrolyzed.

In view of these results, we felt a more exhaustive solvent extraction of flue-cured tobacco, and pyrolysis of the fractions obtained thereby, might lend additional insights into precursor-product relationships in the smoking process. For reference, the reported composition of flue-cured tobacco is presented in Table IV.

Table III. Pyrolytic Production of Benzo(a)pyrene (5)

Material Pyrolyzed	Weight	% Leaf	Yield BaP	% Total
Flue-cured tobacco	25.0 g	100	5500 μg	100
Hexane extract	1.4 g	5.6	3350 μg	61
Extracted tobacco	23.6 g	94.4	2100 μg	38

Table IV.   Chemical Composition of Flue-Cured Tobacco

Leaf Fraction	Constituents	% Dry Leaf Weight	
Ash	inorganics	13.5 (21)	9.25 (22)
Crude fibers	cellulose, lignin	11.2	7.34
Carbohydrates	poly & monosaccharides, starch, dextrin	22.5	36.35
Pectins	pectinic acids	8.0	8.48
Organic acids	Krebs cycle acids	12.2	9.96
Ether-solubles	oils, waxes, resins	7.3	6.61
Tannins	polyphenols	2.2	----
Nitrogen compounds	proteins, amino acids, nitrates, alkaloids	15.2	----

It is noted that the leaf consists of a preponderance of carbohydrates, crude fibers, pectins, etc., with smaller amounts of
nitrogenous materials, lipids, and polyphenols. Sequential extraction of flue-cured tobacco with solvents of increasing polarity
yielded the series of extracts presented in Table V. The composition of these extracts was monitored by thin-layer chromatography,
and appropriate colorimetric analyses; constituents listed in the
right-hand column of the table indicate an approximate order of extraction of the various classes of leaf component. Although some
overlapping of components occurred, the initial three extracts,
accounting for approximately 25% of dry leaf weight, essentially
removed all the waxes, oils, sterols, and terpenes; ethanol extracted the brown pigments and nicotine salts, while the remaining
material was largely carbohydrate.

The individual extracts and the residue were each pyrolyzed
(860°C, N_2) and fractions containing hydrocarbons, phenols, and

Table V.   Sequential Solvent Extraction of Flue-Cured Tobacco (23)

Solvent	% Dry Leaf Weight	Constituents
Skellysolve	7.2	Lipids, including
Chloroform	2.1	Waxes, oils,
Acetone	17.5	Sterols and terpenes.
Ethanol	12.0	Pigment, nicotine salts
Methanol	7.0	Krebs cycle acids
Water	10.7	Carbohydrates,
Residue	43.5	Cellulose, lignin, protein

Table VI.   Aromatic Hydrocarbons in Pyrolysates
Of Tobacco and Tobacco Extracts (23)

Benzene	Acenaphthylene
Toluene	Acenaphthene
Xylenes	Anthracene
Ethylbenzene	Phenanthrene
Styrene	Alkyl-Anthracenes
Indene	Alkyl-Phenanthrenes
Naphthalene	Fluoranthene
Alkyl-Naphthalenes	Pyrene
Biphenyl	Chrysene
Fluorene	Benzo(a)pyrene

nitrogen-containing compounds isolated, and compared with corresponding fractions obtained from pyrolysis of flue-cured tobacco. In all cases, major products obtained were qualitatively similar to those in the tobacco pyrolysate, but with significant quantitative variations. Products identified in the hydrocarbon fraction of the pyrolysates are listed in Table VI. Quantities of these aromatic hydrocarbons in the extract pyrolysates are presented in Table VII. The skellysolve through acetone extracts, removing approximately 25% of dry leaf weight and essentially all leaf lipids, accounted for 72% of the total aromatic hydrocarbon and 92% of the total benzo(a)pyrene content of a tobacco pyrolysate. The remaining 75% of the leaf is seen to produce relatively low levels of aromatic hydrocarbons on pyrolysis. Interestingly, the cumula-

Table VII.   Contributions of Leaf Extracts to Levels of Aromatic
Hydrocarbons in Tobacco Pyrolysate

Extract	% Dry Leaf Weight	% Contribution to Tobacco Pyrolysate	
		Total A.H.	B(a)P
Skellysolve	7.2	33.33	26.80
Chloroform	2.1	4.53	7.22
Acetone	17.5	35.37	59.28
Ethanol	12.0	2.56	< 1
Methanol	7.0	1.61	< 1
Water	10.7	2.59	< 1
Residue	43.5	16.55	7.22
Total	100	96.54	100.52

tive levels of aromatic hydrocarbons obtained by pyrolysis of the individual extracts and residue accounts for almost 97% of that obtained on pyrolysis of whole tobacco, indicating that, at least under these pyrolytic conditions, synergistic effects are negligible.

We shall next examine results of this experiment with regard to the pyrolytic yields of volatile phenols. Some background material regarding past studies of phenol precursors is presented in Table VIII.

In 1939, Wenusch first suggested quinic acid, a moiety of the chlorogenic acid ester, as a pyrolytic source of phenols. Zane and Wender pyrolyzed chlorogenic acid, rutin, and quercetin and identified a number of dihydroxy-benzene derivatives. Extensive investigations into the source of cigarette smoke phenols were performed by a group headed by A. W. Spears (cf. ref. 7). This group extracted flue-cured tobacco with hexane (7% yield) and 75% ethanol (47% yield); subsequent pyrolysis of these extracts (685°C, N_2) resulted in data indicating that the ethanol extract was a considerably more potent phenol precursor than either the hexane extract or whole tobacco. Assuming the ethanol extract to consist essentially of carbohydrate, Spears utilized C^{14}-labeled glucose in cigarettes to estimate that 41% of the smoke phenols are attributable to pyrolysis of leaf carbohydrate (this figure assumes that glucose is typical of leaf carbohydrate and that 55% of leaf is carbohydrate). Subsequent experiments performed at USDA (Table IX) indicated that a wide range of potential for pyrolytic production of phenol exists among various leaf constituents.

Table VIII.   Pyrolytic Precursors of Smoke Phenols

Precursor	Reference(s)
Quinic acid	Wenusch (3)
Chlorogenic acid	Zane and Wender (24)
Carbohydrates	Bell et al. (7)
	Schlotzhauer et al. (6)
Brown pigments	Schlotzhauer et al. (6)
Lignin	Kato et al. (25)
	Schlotzhauer et al. (6)
Organic acids	Schmeltz et al. (26)

Table IX. Yields of Phenol by Pyrolysis of Leaf Constituents (6)

Constituent	mg. Phenol/ 100 g. Pyrolyzed	Relative Yield
Brown pigments	174	21.75
Lignin	104	13.00
Glucose	39	4.87
Polygalacturonic acid	29	3.62
Glucuronic acid	27	3.37
Cellulose	8	1.00

It is evident from the preceding data that carbohydrates are
relatively poor precursors of phenol in comparison to lignin and
brown pigments. The latter, which have been characterized by
Wright et al. (27) and Chortyk et al. (28), among others, as iron-
protein-chlorogenic acid-rutin complexes, produced approximately
4.5 times the yield of phenol pyrolytically than did glucose
(Spears' typical carbohydrate) and more than 21 times the yield
obtained from cellulose. These past observations add considerable
insight toward interpretation of the data obtained from pyrolysis
of the various tobacco extracts. Components identified in the
pyrolysates of tobacco, and the tobacco extracts, included phenol,
the isomeric cresols, and lesser amounts of xylenols. Quantitative
analyses of the pyrolysates are presented in Table X. The ethanol
extract and the tobacco residue account for more than 80% of the
volatile phenol content of a tobacco pyrolysate. Interestingly,
the ethanol extract, although only about one-fourth the weight of
the residue, contributes nearly as high a proportion of these
phenols as the latter. It is suggested that the 3 to 5% of brown
pigments of leaf, concentrated in the ethanol extract, being
considerably more potent precursors of phenols than the carbohydrates remaining in the leaf residue, account for this observation.

Table X. Contributions of Leaf Extracts to the Levels
Of Volatile Phenols in Tobacco Pyrolysate (23)

Extract	% Dry Leaf Weight	% Contribution to Phenols in Tobacco Pyrolysate
Skellysolve	7.2	2.68
Chloroform	2.1	0.89
Acetone	17.5	8.03
Ethanol	12.0	38.39
Methanol	7.0	3.12
Water	10.7	2.68
Residue	43.5	43.75
Total	100	99.54

The final class of compounds to be examined in this study are the nitrogen-containing components. Because of the toxicity and high content of nicotine in cigarette smoke condensates, biological testing of such condensates and fractions thereof must be conducted on a nicotine-free basis. This is especially true for the basic fraction of smoke condensate, which Wynder and Wright (29) and Wynder and Hoffmann (30) found to be weakly tumorigenic and low in tumor-promoting activity. The presence of the carcinogenic N-heterocyclic hydrocarbons, the dibenz-acridines, in cigarette smoke has been observed by Van Duuren et al. (31). That N-heterocyclic hydrocarbons can arise from pyrolysis of nicotine has long been noted; Jarboe and Rosene (32) have reported the major pyrolysis products of nicotine to consist of a series of pyridine bases, preferably 3-substituted, plus quinoline and isoquinoline, nitrogen-containing analogs of naphthalene. In addition to the tobacco alkaloids, a variety of nitrogenous leaf components can give rise to N-heterocyclic compounds (33).

Of particular interest to smoking and health researchers is the suggestion that the secondary amines of tobacco leaf may give rise through thermal reactions to the potent tumor initiating N-nitrosamines. Recently, Hoffmann and Vais (34) have reported the isolation from cigarette smoke of five N-nitrosamines (as the corresponding hydrazones); however, evidence on the source and mode of formation of these compounds in smoke is currently incomplete.

Pyrolysis of tobacco and the various extracts give rise to pyridine, picolines, 3-ethyl-pyridine, 3-vinylpyridine, 3-cyano-pyridine (nicotinonitrile), quinolines, and benzoquinolines. In Table XI, nicotinonitrile has been quantitated since this product of thermal degradation of nicotine was a major component in all pyrolysates examined.

Table XI. Contributions of Leaf Extracts to Levels Of Nicotinonitrile in Tobacco Pyrolysate (23)

Extract	% Dry Leaf Weight	% Contribution to Tobacco Pyrolysate
Skellysolve	7.2	10.03
Chloroform	2.1	8.49
Acetone	17.5	1.93
Ethanol	12.0	30.12
Methanol	7.0	6.95
Water	10.7	1
Residue	43.5	3.09
Total	100	60.61

Although only 60% of the nicotinonitrile in a tobacco pyrolysate could be accounted for by examining the pyrolysates of the fractions listed in Table XI, half of this total was concentrated in the pyrolysate of the ethanol extract. Nicotine salts are generally extractable with alcohol; moreover, Dymicky et al. (35, 36) have implicated alkaloids and simple pyridine bases in the structural makeup of the brown pigments, which are also extractable with ethanol. The protein moieties of these pigments would also contribute to a concentration of basic products in the pyrolysate of the ethanol extract.

In summary, we have discussed the role of pyrolysis in smoking and health research; we have reviewed some of the findings of past investigations, and presented representative data obtained by pyrolysis of extracts from sequential extraction of flue-cured tobacco. Data indicated that the leaf lipids preferentially contribute to the levels of aromatic hydrocarbons, especially benzo-(a)pyrene, obtained in tobacco pyrolysis. Similar preference for pyrolytic production of volatile phenols is shown by the brown pigments and leaf carbohydrates. Bases, including N-heterocyclic compounds, largely arise from pyrolysis of nicotine of both the bound and unbound variety. This discussion on pyrolysis was limited in scope, and unfortunately, could not include all such contributions to tobacco smoke as have been made over the years by many investigators, but, hopefully, will provide a broader view as to the value of pyrolysis toward better understanding of the complex relationships between tobacco leaf and tobacco smoke.

References

1. Stedman, R. L. Chem. Rev. 68: 153-207 (1968).
2. Gilbert, J. A. S., and Lindsey, A. J. Brit. J. Cancer 11: 398 (1957).
3. Wenusch, A. Oesterr. Chem. Ztg. 42: 226-231 (1939).
4. Wynder, E. L., Wright, G. F., and Lam, J. Cancer 11: 1140-1148 (1958).
5. Schlotzhauer, W. S., and Schmeltz, I. Beitr. Z. Tabakforsch. 4: 176-181 (1968).
6. Schlotzhauer, W. S., Schmeltz, I., and Hickey, L. C. Tob. Sci. 11: 31-34 (1967).
7. Bell, J. H., Saunders, A. O., and Spears, A. W. Tob. Sci. 10: 138-142 (1966).
8. Newsome, J. R., and Keith, C. H. Tob. Sci. 9: 75-79 (1965).
9. Touey, G. P., and Mumpower, R. C. Tob. Sci. 1: 33-37 (1957).
10. Lam, J. Acta. Pathol. Microbiol. Scand. 36: 503-510 (1955).
11. Benner, J. R., Burton, H. R., and Burdick, D. Beitr. Z. Tabakforsch. 5: 74-79 (1969).

12. Ayres, C. I., and Thornton, R. E. Abstracts, 19th Tobacco Chemists' Research Conference, Lexington, Ky. (1965).
13. Cooper, R. L., and Lindsey, A. J. Chem. Ind. (London) 1205 (1953).
14. Wynder, E. L., and Hoffmann, D. Cancer 12: 1194-1199 (1959).
15. Badger, G. M., Donnelly, J. K., and Spotswoods, T. McL., Australian J. Chem. 18:1249-1266 (1965).
16. Schlotzhauer, W. S., and Schmeltz, I. Beitr. Z. Tabakforsch. 5: 5-8 (1969).
17. Gil-Av, E., and Shabati, J. Nature 197: 1065-1066 (1963).
18. Oro, J., Han, J., and Zlatkis, A. Anal. Chem. 39: 27-32 (1967).
19. Swain, A. P., Rusaniwskyj, W., and Stedman, R. L. Chem. Ind. 14: 435-436 (1961).
20. Rayburn, C. H., Wartman, W. B., and Pederson, P. M. Science 128: 1344-1345 (1958).
21. Frankenburg, W. G. Advan. Enzymol. 6: 309-387 (1946).
22. Bacon, C. W., Wenger, R., and Bullock, J. F. Ind. Eng. Chem. 44: 292-296 (1952).
23. Schlotzhauer, W. S., Chortyk, O. T., Higman, H. C., and Schmeltz, I. Tob. Sci. 13: 153-155 (1969).
24. Zane, A., and Wender, S. H. Tob. Sci. 7: 21-23 (1963).
25. Kato, K., Sukai, F., and Nakabota, T. Japan Monopoly Corp. Cent. Res. Inst. Scient. Papers 107: 171-175 (1965).
26. Schmeltz, I., Hickey, L. C., and Schlotzhauer, W. S. Tob. Sci. 11: 52-53 (1967).
27. Wright, H. E., Burton, W. W., and Berry, R. C. Phytochem. 3: 525 (1964).
28. Chortyk, O. T., Schlotzhauer, W. S., and Stedman, R. L. Beitr. Z. Tabakforsch. 3: 421-428 (1966).
29. Wynder, E. L., and Wright, G. F. Cancer 10: 1201-1205 (1957).
30. Wynder, E. L., and Hoffmann, D. (Eds.) Tobacco and Tobacco Smoke. Academic Press, New York (1967).
31. Van Duuren, B. L., Bilbao, J. A., and Joseph, C. A. J. Natl. Cancer Inst. 25: 53-61 (1960).
32. Jarboe, C. H., and Rosene, C. J. J. Chem. Soc. 2455-2458 (1961).
33. Schmeltz, I., Schlotzhauer, W. S., and Higman, E. B. Beitr. Z. Tabakforsch. (In press).
34. Hoffmann, D., and Vais, J. Abstracts, 25th Tobacco Chemists' Research Conference, Louisville, Ky. (1971).
35. Dymicky, M., and Stedman, R. L. Phytochem. 6: 1025-1031 (1967).
36. Dymicky, M., Chortyk, O. T., and Stedman, R. L. Tob. Sci. 11: 42-44 (1967).

RECENT ADVANCES IN KNOWLEDGE OF THE CHEMICAL COMPOSITION

OF TOBACCO SMOKE

Georg B. Neurath

2000 Hamburg 56, Hexentwiete 32, West Germany

Since the last comprehensive report on the chemical composition of tobacco and tobacco smoke was published by R. L. Stedman
in 1968, more than 300 additional constituents have been identified, bringing the total number of known compounds in tobacco
smoke to more than 1150.

During this time the main interest has obviously been focused
on a more detailed elucidation of the composition of some important
fractions of the whole smoke and their behavior during the smoking
process. Thus, sophisticated analyses of the polycyclic aromatic
hydrocarbons have been carried out by Rathkamp and Hoffmann,
Carugno, and Grimmer. The fraction of the aliphatic and aromatic
nitro compounds has been elucidated by Hoffmann. Woodman, Owen,
and Westcott have studied the chemical composition of various particle size fractions of the smoke aerosol. Pyrolysis studies have
been carried out by Schmeltz, and some interest has been spent on
the isolation and identification of the so-called "semivolatiles"
of tobacco smoke by Grob and our group. But before dealing with
this matter, which will take the main part of this paper, I should
like to give a short summary and some comments on the recently
identified constituents of several classes of compounds.

The almost complete variety of lower hydrocarbons in tobacco
smoke was further enlarged by the identification of dimethylcyclo-
propanes by Bartle and Novotny. The other newly identified aliphatic hydrocarbons consist mainly of homologues and branched
isomers of known compounds. Among the aromatic hydrocarbons,
methylstyrenes have been identified by Glock et al. Other papers
describe methyl derivatives of acenaphthylene (Entwistle and

Johnstone), fluorene and fluoranthene (Rathkamp and Hoffmann), indanes, indenes, and tetrahydronaphthalenes (Neurath, Gewe, and Wichern).

Some of the carbonyl compounds which have recently been identified will be treated later in this paper, but the identification of a series of alkyl fluorenones by Bell, Ireland, and Spears as well as by Testa may be mentioned here.

Further alkyl phenols including 2-butyl- and 2-isobutylphenol have been found by Miller, Strange, and Stedman. Methyl-, ethyl-, n-propyl-, and isopropylpyrocatechol could be identified by Kallianos, Means, and Mold.

The whole scale of aliphatic and aromatic amines seems now to have been discovered after two further publications by the Vienna group of Pailer, Kuhn, et al. Their papers also cover alkyl pyrrolidines and alkyl piperidines.

The knowledge of alkyl indoles and alkyl carbazoles has been enlarged by Hoffmann and Rathkamp. A number of pyridines, pyrazines, and quinolines has been added to the list of the tobacco smoke constituents during the last few years (Neurath and Dünger).

Also greatly enlarged was the number of identified aromatic and heterocyclic cyanides by Miller, Strange, and Stedman and our group.

Some benzo- and dibenzofurans have been found, and some very interesting furyl compounds, i.e., difuryl methanes and furyl pyrazines, that we will treat later on in this paper.

Two papers of Hoffmann and Rathkamp dealt with the occurrence of nitroalkanes and nitroalkyl benzenes in tobacco smoke.

Leach and Alford reported on the occurrence of disaccharides and erythritol in tobacco smoke. And Frederickson has been able to identy β-amyrin esters of octacosanoic and hentriacontanoic acids in the smoke.

Last but not least, I should like to mention the list of inorganic ions added by Nadkarni, Ehmann, and Burdick, including gold and silver as well as selenium, bromine, scandium, lanthanum, and cobalt.

But now I should like to treat a special chapter of recent investigations on tobacco smoke dealing with the so-called "semivolatiles." The term "semivolatile compounds of tobacco smoke" was used for the first time by Williamson, Graham, and Allman in

1965 in an investigation of filter selectivity problems for "those substances which are retained on a Cambridge filter but which can be volatilized from it."

The possibility of selectively removing substances from tobacco smoke is as far as is presently known restricted to compounds showing measurable vapor pressure at ambient temperatures - in other words, to the semivolatile compounds.

There are two more reasons which stress the importance of this group of chemically very different smoke constituents with only one property in common - their boiling points:

1. Due to their volatility they might be considered as possibly aroma-bearing or -contributing individuals.

2. The semivolatiles can easily be identified compared with nonvolatile compounds. Thus, one may get hints from the semivolatiles at other compounds or classes of compounds which are not volatile.

The isolation of semivolatile compounds can be achieved in various ways. For instance Williamson, Graham, and Allman obtained semivolatiles by heating the tobacco smoke condensate being adsorbed on Cambridge filters. They liberated the semivolatiles from the filters using an argon flow in an oven kept at 180°C to 190°C. Grob obtained the semivolatiles by heating activated charcoal filters loaded with tobacco smoke condensate to 120°C, whereas Levins and Ikeda put their Cambridge filters loaded with condensate directly into the injection port of a gas chromatograph.

We decided to use tobacco smoke condensate from dry ice-acetone traps and to isolate the semivolatiles by a simple steam distillation, regarding the possibility of artifact formation by this procedure as negligible. The steam distillate was separated by column chromatography on silica gel into six fractions. Identification of components was based on combined gas chromatography--mass spectrometry, and on comparison of the spectra obtained with those of authentic compounds. The total yield of semivolatiles by this procedure was 7.46% of the original smoke condensate. Inasmuch as the experimental procedure used will be published elsewhere (Beiträge zur Tabakforschung), only a brief description will be presented here.

We started with 366 g smoke condensate and got 27.6 g steam distillate, which were separated on silica gel into six fractions shown in Table 1. The elution was discontinued after the ethyl acetate fraction was collected. The total eluate off the column amounted to 68% of the steam distillate.

The gas-chromatographic separation was performed on a 100-m Emulphor capillary column with helium as carrier gas. A temperature program was run from 90°C to 180°C at 1.25°C per minute increase. The exit part was split in a 5:1 ratio between a flame ionization detector and the mass spectrometer.

The identification of all compounds listed was accomplished by comparison of their mass spectra with the spectra of authentic, synthetic samples, and by comparison of the retention times, of course. In some cases, in particular with alkylated aromatic hydrocarbons and other alkylated aromatic compounds, it was not possible to determine the exact nature of possible isomers.

Fraction 1, eluted with n-hexane (Figure 1), as expected comprises aliphatic saturated and Δ^3- to Δ^4-unsaturated hydrocarbons, and dimethyl naphthalene (19) as the only identified aromatic hydrocarbon in this fraction. Farnesenes (14) and neophytadiene (23) predominate in this fraction. The occurrence of 2,6,10,14-tetramethyl-pentadecene-6 (20) may be mentioned. The complete list of substances identified in this fraction is given in Table 2.

Fraction 2, eluted with benzene (Figure 2, Table 3), consists chiefly of methylated benzenes, naphthalenes, indanes, indenes, and benzofurans. Thus, peak 4 is indene, peaks 10 are two isomers of methyl benzofuran, 11 three isomers of methyl indene, 13 three isomers of dimethyl benzofuran, 14 two isomers of dimethyl indane; among them 15 is naphthalene followed by 16 five isomers of dimethyl indene. The area 18 to 26 is covered by tri- and tetramethyl derivatives of benzofuran, indane, and indene. Peak 21 is 2-methyl-naphthalene. The fraction ceases with three-ring compounds: thus, 30 is acenaphthene, 32 a methyl derivative of it; 33 is acenaphthylene, 34 dibenzofuran. 37 is fluorene, and at the end 38 is phenanthrene. These compounds are overlapped by eight isomers of trimethylnaphthalene. Peaks 9, 12, and 17 represent difuryl methanes, which have not been previously described to occur in tobacco smoke.

Fraction 3, also eluted by benzene but differing in color at the column (Figure 3, Table 4), contains a lot of cyanides as main constituents: thus, 13 is benzonitrile; 17, 18, and 19 are the three tolunitriles, followed by 21 benzyl cyanide and four dimethyl-benzonitriles. Peak 33 is cinnamonitrile. Also, peak 6 is a nitrile, 5-methyl-2-cyano-furan. The whole fraction is overlapped by methylnaphthalenes, -indanes, and -indenes, which is not surprising according to the conditions. An interesting compound is peak 26, representing N-furfurylpyrrol, a compound, which is not known in the natural scenery. The fraction ends in 34 indole and 36 skatole, with some methyl derivatives.

From the quantitative point of view, fraction 4 (Figure 4, Table 5), eluted with diisopropyl ether, is the phenol fraction. It is represented by peaks 18 and 19, phenol and cresols, but it contains a variety of cycloaliphatic ketones, furan compounds, and phenones, some of them not previously reported in tobacco smoke. Thus, 1 is cyclopentanone; 3 is 3-formylfuran, the isomer of furfural (4). Peak 6 is acetylfuran; 7 is furfuryl acetate, and peak 8 is 5-methylfurfural. 10 is α-isophorone and peak 13 3,5,5-trimethylcyclohexene-2-dione-1,4. 12 is acetophenone; 15 and 16 are o- and p-methylacetophenone, and 14 is propiophenone. Last but not least, 17 is solanone.

Fraction 5 (Figure 5, Table 6), the ether fraction, to some extent continues fraction 4. Thus, peak 2 is cyclopentene-2-one; 3 and 12 are its 2- and 3-methyl derivatives; 6 is cyclohexene-2-one and peak 8 represents acetylfuran. But 5 and 11, both of them large peaks, represent a new class of compounds: acetoxypropanone and acetoxybutanone-2. Another in a qualitative sense predominating class of compounds in this fraction are the pyrazines, represented by peaks 4, ethylpyrazine; 7, which is 5- or 6-methyl-2-ethyl-pyrazine; 10 2,5-, and 13 2,6-diethylpyrazine. 16 is tetramethylpyrazine. 25 is the remarkable furylpyrazine (furyl-2'-pyrazine), followed by its methyl and dimethyl derivatives 26, 28, and 30. The large peak 18 is Δ^1-tetrahydrobenzaldehyde. 19 is vinylpyridine and 21, 22, and 24 represent cyanopyridines including the methyl and dimethyl derivatives. The last identified peak in this fraction, 32, is phenylpyridine.

The ethylacetate fraction 6 (Figure 6, Table 7) contains a further lot of pyrazines. Thus, peak 1 is the unsubstituted compound, peak 2 is methylpyrazine; 4, 5, and 6 are, respectively, the 2,6-, 2,5-, and 2,3-dimethyl derivatives. 7 and 8 represent methyl-ethylpryazines (2,6- and 2,5-); methylpropyl- and methylisopropyl-pyrazine are also present (11 and 12). The large peak 13 is tetramethylpyrazine. Pyridines occur as peak 10 (methylethyl derivative) and peaks 14 (2-methyl-5-isopropyl-), 15 (3-propyl-), and 16 (2,4-dimethyl-5-isopropylpyridine). Peak 17 is butyro-lactone. Peaks 20, 22, and 24 are 3-acetyl-, 2-methyl-5-acetyl-, and 3-propionylpyridine. Quinoline and its methyl and dimethyl derivatives are represented by peaks 23, 25, and 26.

Although there is a real difference between the methods used for this study and that utilized by Grob, i.e., by-passing solvent without a preceding column chromatography, it may be attractive to reconsider the results of both papers. Grob observed 350 peaks, compared to 320 different compounds indicated by our mass spectra. Of the 133 compounds identified by Grob and the 215 added by this paper, 57 constituents are identified by both groups. Whereas Grob identified 76 components we did not detect, there is a group

of 158 compounds not mentioned by Grob. Thus, the overall result
could possibly be the identification of more than 290 constituents
out of 320 to 350 observed in all.

Table 1

Column Chromatography on Silica Gel

Fraction No.	Eluted with	Quantity (l)	Eluate (g)	Part of the steam distillate (%)
1	n-Hexane	2	1.72	6.30
2	Benzene	0.2	0.782	2.86
3	Benzene	1.8	0.943	3.45
4	Diisopropyl ether	2	9.657	35.37
5	Ether	2	4.806	17.60
6	Ethyl acetate	2	0.627	2.30
			18.535	67.88 %

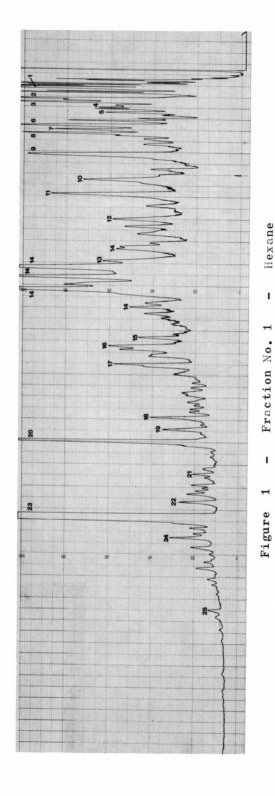

Figure 1   -   Fraction No. 1   -   Hexane

Table 2	–	Fraction No. 1	–	n-Hexane

Peak No.	Compound	First report in tobacco smoke
1	Limonene	
2	$C_{12}H_{26}$	
3	Δ^2-$C_{12}H_{24}$	
4	$C_{13}H_{26}$	
5	$C_{14}H_{30}$	
6	n-Tridecane	
7	Δ^2-$C_{14}H_{28}$	
8	n-Tridecene-2	
9	$C_{15}H_{30}$	
10	n-Tetradecane	
11	n-Tetradecene-2	
12	$C_{16}H_{34}$	
13	n-Pentadecane	
14	5 Isomers of farnesene $C_{15}H_{24}$	
15	n-Hexadecane	
16	$C_{15}H_{26}$	
17	$C_{17}H_{34}$	
18	n-Heptadecane	
19	Dimethylnaphthalene	
20	2,6,10,14-Tetramethylpentadecene-6	+
21	8-Methylheptadecane	+
22	3-Methylene-7,11,15-trimethylhexadecane	+
23	Neophytadiene (3-Methylene-7,11,15-trimethylhexadecene-1)	
24	Phytadiene (3,7,11,15-Tetramethylhexadecadiene-1,3)	
25	4-Methylene-8,12,16-trimethylheptadecene-2	+

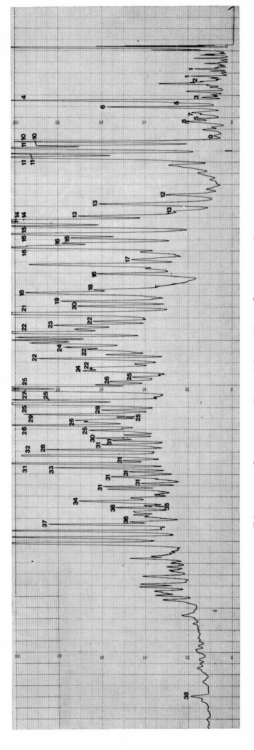

Figure 2 — Fraction No. 2 — Benzene

Table 3	—	Fraction No. 2	—	Benzene

Peak No.	Compound	First report in tobacco smoke
1	2 Isomers of trimethylbenzene	
2	N-(3-Methylbutyl-1-)pyrrole	+
3	Benzofuran	
4	Indene	
5	2 Isomers of tetramethylbenzene	
6	2-Methyl-5-hexen-3'-yl-furan	+
7	Methylhexenyl furan	+
8	C_4-Benzene	
9	Di-α-furylmethane	+
10	2 Isomers of methylbenzofuran	
11	3 Isomers of methylindene	
12	5-Methylfuryl-2-furyl-2'-methane	+
13	3 Isomers of dimethylbenzofuran	
14	2 Isomers of dimethylindane	
15	Naphthalene	
16	5 Isomers of dimethylindene	
17	5,5'-Dimethyl-di-α-furylmethane	+
18	2 Isomers of trimethylbenzofuran	+
19	Ethylindene	
20	Trimethylindane	+
21	2-Methylnaphthalene	
22	5 Isomers of trimethylindene	+
23	1-Methylnaphthalene	
24	2 Isomers of tetramethylbenzofuran	+
25	6 Isomers of dimethylnaphthalene	
26	3 Isomers of tetramethylindene	+
27	Diphenyl	
28	2 Isomers of methyldiphenyl	+
29	C_4-Naphthalene	
30	Acenaphthene	
31	8 Isomers of trimethylnaphthalene	
32	Methylacenaphthene	+
33	Acenaphthylene	
34	Dibenzofuran	
35	Tetramethylnaphthalene	+
36	2 Isomers of methyldibenzofuran	
37	Fluorene	
38	Phenanthrene	

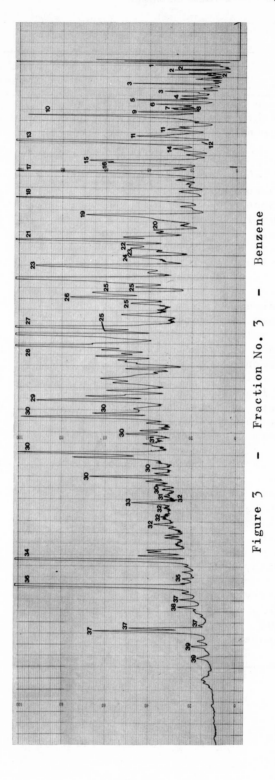

Figure 3 — Fraction No. 3 — Benzene

Peak No.	Compound	First report in tobacco smoke
Table 4 — Fraction No. 3 — Benzene		
1	Toluene	
2	o-, m-, p-Xylenes	
3	2 Isomers of trimethylbenzene	
4	Tetramethylbenzene	
5	C_3-Benzene	
6	5-Methyl-2-cyanofuran	+
7	Methoxytoluene	
8	4-Isopropenyltoluene	
9	Indene	
10	Benzaldehyde	
11	2 Isomers of methylindane	
12	Dimethylindane	
13	Benzonitrile	
14	Coumaran	
15	Methylindene	
16	Methylbenzaldehyde	
17-19	o-, m-, p-Tolunitriles	+
20	Methylcoumaran	+
21	Benzylcyanide	
22	Propiophenone	
23	2 Isomers of dimethylindene	
24	Naphthalene	
25	4 Isomers of dimethylbenzonitrile	+
26	N-(Furfuryl-2'-)pyrrole	+
27	2-Methylnaphthalene	
28	1-Methylnaphthalene	
29	Xylenol	
30	7 Isomers of dimethylnaphthalene	
31	Trimethylphenol	
32	4 Isomers of trimethylnaphthalene	
33	Cinnamonitrile	
34	Indole	
35	Methyl palmitate	
36	Skatole	
37	4 Isomers of dimethylindole	
38	Methylindole	
39	2 Isomers of trimethylindole	

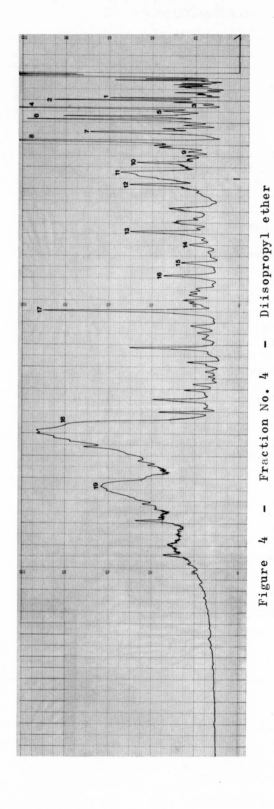

Figure 4 — Fraction No. 4 — Diisopropyl ether

Table 5	–	Fraction No. 4	–	Diisopropyl ether
Peak No.	**Compound**			**First report in tobacco smoke**
1	Cyclopentanone			
2	$C_7H_{10}O$			
3	3-Formylfuran			+
4	Furfural			
5	$C_8H_{12}O$			
6	2-Acetylfuran			
7	Furfuryl acetate			+
8	5-Methylfurfural			
9	5-Methyl-2-acetylfuran			+
10	α-Isophorone			+
11	Furfuryl alcohol			
12	Acetophenone			
13	3,5,5-Trimethylcyclohexene-2-dione-1,4			+
14	Propiophenone			
15	o-Methylacetophenone			+
16	p-Methylacetophenone			+
17	Solanone			
18	Phenol			
19	Cresol			

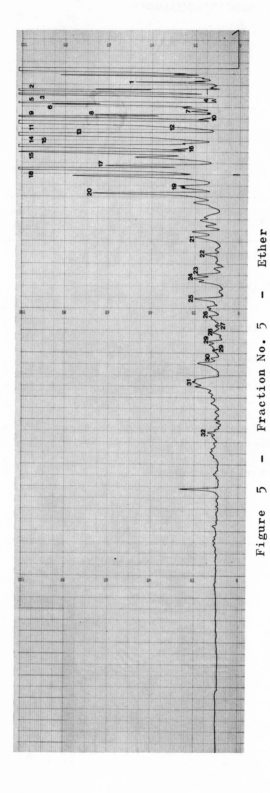

Figure   5   –   Fraction No. 5   –   Ether

Table 6	—	Fraction No. 5	—	Ether

Peak No.	Compound	First report in tobacco smoke
1	Methyltetrahydrofuranone	+
2	Cyclopenten-2-one	+
3	2-Methylcyclopenten-2-one	+
4	Ethylpyrazine	+
5	Acetoxypropanone	+
6	Cyclohexen-2-one	+
7	5- or 6-Methyl-2-ethylpyrazine	
8	2-Acetylfuran	
9	$C_7H_{10}O$	
10	2,5-Diethylpyrazine	+
11	1-Acetoxybutanone-2	+
12	3-Methylcyclopenten-2-one	+
13	2,6-Diethylpyrazine	
14	$C_7H_{10}O$	
15	2 Isomers of $C_8H_{12}O$	
16	Tetramethylpyrazine	
17	$C_6H_8O_2$	
18	Δ^1-Tetrahydrobenzaldehyde	+
19	3-Vinylpyridine	
20	$C_5H_6O_2$	
21	3-Cyanopyridine	
22	Methylcyanopyridine	+
23	Quinoline	
24	Dimethylcyanopyridine	+
25	Furyl-2'-pyrazine	+
26	6-Methyl-2-(furyl-2'-)pyrazine	+
27	Methylquinoline	
28	5-Methyl-2-(furyl-2'-)pyrazine	
29	2 Isomers of dimethylquinoline	
30	5,6-Dimethyl-2-(furyl-2'-)pyrazine	+
31	2-Acetylpyrrole	+
32	3-Phenylpyridine	

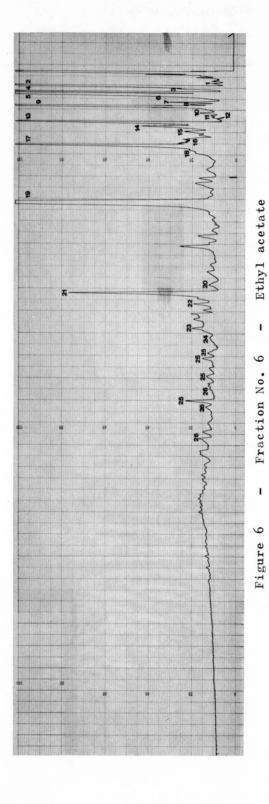

Figure 6   —   Fraction No. 6   —   Ethyl acetate

Table 7 - Fraction No. 6		Ethyl acetate
Peak No.	Compound	First report in tobacco smoke
1	Pyrazine	
2	Methylpyrazine	
3	Dimethylpyridine	
4	2,6-Dimethylpyrazine	
5	2,5-Dimethylpyrazine	
6	2,3-Dimethylpyrazine	
7	2-Methyl-6-ethylpyrazine	
8	2-Methyl-5-ethylpyrazine	+
9	Trimethylpyrazine	
10	Methyl-ethylpyridine	+
11	Methyl-isopropylpyrazine	+
12	Methyl-propylpyrazine	+
13	Tetramethylpyrazine	
14	2-Methyl-5-isopropylpyridine	+
15	3-Propylpyridine	+
16	2,4-Dimethyl-5-isopropylpyridine	+
17	Butyrolactone	+
18	3-Vinylpyridine	
19	$C_4H_4O_2$	
20	3-Acetylpyridine	
21	$C_5H_6O_2$	
22	2-Methyl-5-acetylpyridine	+
23	Quinoline	
24	3-Propionylpyridine	
25	4 Isomers of methylquinoline	
26	3 Isomers of dimethylquinoline	

Literature

K.D.Bartle and M.Novotný
Beiträge Tabakforsch. 5 (1970) 215

J.H.Bell, S.Ireland, and A.W.Spears
Anal.Chem. 41 (1969) 310

N.Carugno and S.Rossi
J.Gaschrom. 5 (1967) 103

I.D.Entwistle and R.A.W.Johnstone
J.Chem.Soc. (1968) 1818

J.D.Frederickson
20th Tob.Chem.Res.Conf., Winston-Salem, N.C. (1966)

G.Grimmer
5th Intern.Tob.Sci.Congr., Hamburg, West Germany (1970)

K.Grob and J.A.Völlmin
Beiträge Tabakforsch. 5 (1969) 52

K.Grob
5th Intern.Tob.Sci.Congr., Hamburg, West Germany (1970)

D.Hoffmann and G.Rathkamp
Beiträge Tabakforsch. 4 (1968) 124

D.Hoffmann and G.Rathkamp
Anal.Chem. 42 (1970) 1643

D.Hoffmann, G.Rathkamp, and H.Woziwodzki
Beiträge Tabakforsch. 4 (1968) 253

A.G.Kallianos, R.E.Means, and J.D.Mold
21st Tob.Chem.Res.Conf., Durham, N.C. (1967)

J.T.Leach and E.D.Alford
22nd Tob.Chem.Res.Conf., Richmond, Va. (1968)

R.L.Miller, E.D.Strange, and R.L.Stedman
24th Tob.Chem.Res.Conf., Montreal, Canada (1970)

R.A.Nadkarni, W.D.Ehmann, and D.Burdick
Tob.Sci. 14 (1970)

G.Neurath and M. Dünger
Beiträge Tabakforsch. 5 (1969) 1

G.Neurath and M.Dünger
      5th Intern.Tob.Sci.Congr., Hamburg, West Germany (1970)

M.Dünger, and I. Küstermann
      Beiträge Tabakforsch. 6 (1971) 12

G.Neurath, J.Gewe, and H.Wichern
      Beiträge Tabakforsch. 4 (1968) 247

W.C.Owen, D.T.Westcott, and G.R.Wood
      23rd Tob.Chem.Res.Conf., Philadelphia, Penn. (1969)

M.Pailer, W.J.Hübsch, and H.Kuhn
      Fachl.Mitt.Österr.Tabakregie No.7 (1967) 109

M.Pailer, J.Völlmin, C.Karnincic, and H.Kuhn
      Fachl.Mitt.Österr.Tabakregie No.10 (1969) 165

G.Rathkamp and D.Hoffmann
      24th Tob.Chem.Res.Conf., Montreal, Canada (1970)

I.Schmeltz
      5th Intern.Tob.Sci.Congr., Hamburg, West-Germany (1970)

R.L.Stedman
      Chem.Revs. 68 (1968) 153

P.Testa
      Ann.Direct.Etudes Equipment SEITA 4 (1966) 117

J.T.Williamson, J.F.Graham, and D.R.Allman
      Beiträge Tabakforsch. 3 (1965) 233

FRACTIONATION OF TOBACCO SMOKE CONDENSATE FOR CHEMICAL COMPOSITION STUDIES

William J. Chamberlain[1] and R. L. Stedman[2]

[1]Richard B. Russell Agricultural Research Center, ARS, USDA, Athens, Georgia and [2]Eastern Marketing and Nutrition Laboratory, ARS, USDA, Philadelphia, Pennsylvania.

The evolution of the scheme presently used at the Russell Research Center for the fractionation of smoke condensate will be outlined. This work was carried out for the most part at the Eastern Marketing and Nutrition Laboratory under the direction of Dr. R. L. Stedman. The fractions obtained from the separation procedures are used in both chemical and biological investigations. The results obtained in the biological studies on these fractions will be discussed in another paper by Dr. Bock. Some fractions which were found to be active will be pointed out as a justification for studying the chemistry of the particular fraction and developing further fractionation schemes. This paper is also not intended to be a complete discussion of chemical composition but rather a means of isolating various fractions for chemical studies. Compounds identified in our laboratories will be presented in order to give a general idea of the composition of various fractions.

In our studies we generally start with one kilogram of cigarette smoke condensate (CSC), which is purchased from Roswell Park Memorial Institute in Buffalo, N. Y. The smoke condensate is shipped to us packed in dry ice and is kept frozen until just prior to use. Figure 1 shows the preliminary separation of the condensate into acidic, basic, and neutral fractions. This part of the procedure has not been changed over the years. In our first large-scale fractionation procedure the bases were further separated by adjusting the pH to 11.0 with 12N NaOH and extracting with Et_2O. Components were identified in the Et_2O soluble portion by a GLC technique. These included, several collidines, picolines, and lutidines, 3-ethyl pyridine, pyridine and several alkaloids.

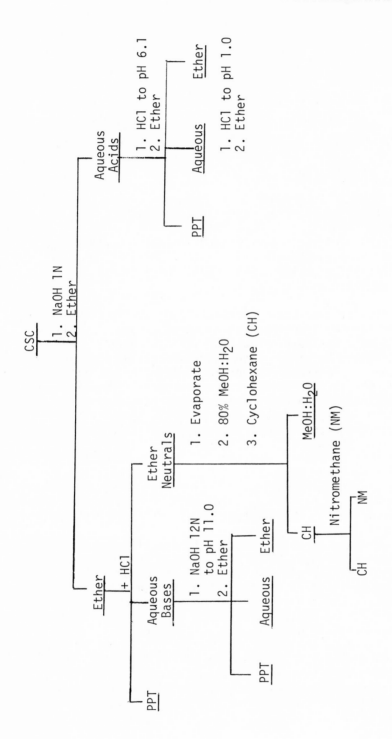

Figure 1.   Scheme 1 – Fractionation of smoke condensate.

The acid fraction was also further separated as follows. The pH is adjusted to 6.1 and extracted with ether to remove the weak acids. The pH is further lowered to 1.0 and the ether-soluble strong acids are removed. This fraction will be discussed later in the paper.

The precipitates shown in Figure 1 are the pigments. There are three subfractions of these: the basic, weakly acidic, and strongly acidic pigments. These have been shown to be similar to the brown pigments of tobacco leaf. The weakly acidic pigment is the largest fraction by weight and represents between 6 and 9 per cent of the condensate weight. Most of the weakly acidic subfraction was found to be nondialyzable and to contain a component having a molecular weight $\gtrless 100,000$, which yielded a silicone, nicotine, and a series of bases on alkaline fusion (1).

We now come to the neutrals. These are divided into three fractions by solvent partitioning between methanol-water, cyclohexane, and nitromethane. This type of separation was first used by Drs. Wynder and Hoffmann (2) to concentrate the polynuclear hydrocarbons, particularly benzo(a)pyrene in the nitromethane fraction. A considerable amount of work has been done in our laboratory to identify some of the components in addition to the polynuclear hydrocarbons in this fraction.

The nitromethane fraction was further separated by silicic acid and alumina column chromatography. These columns are developed by gradually increasing the polarity of the eluting solvent. Some of the compounds which have been identified in the nitromethane fraction were indoles, carbazoles, benzyl benzoate, and myristicin along with polynuclear aromatic hydrocarbons (3). Also in a separate study carried out by R. L. Miller et al., several interesting aryl amines were identified in this fraction (4).

In the large-scale fractionation work done by Dr. Ansel Swain and Mr. Joseph Cooper, an overall recovery average of 90.8% was obtained following nine fractionations of cigarette smoke condensate in 1 kg quantities (5). These fractions were tested at Roswell Park by Dr. Bock and it was found that the weak acid ether-soluble and the three neutral fractions contained the major tumor promoting substances (6). This led us to the development of our second fractionation scheme. Separation of the acids, bases, and neutrals was carried out in the same manner as scheme 1. The basic fraction was not tested in this run. The acids were worked up as shown in Figure 2. In a recent study of the chemical composition of these fractions, Miss Elizabeth Strange has developed a method of separating the long chain fatty acids from the ether-soluble pot fraction by urea occlusion and gas chromatography. By this method she has identified the fatty acids: palmitic, stearic, oleic,

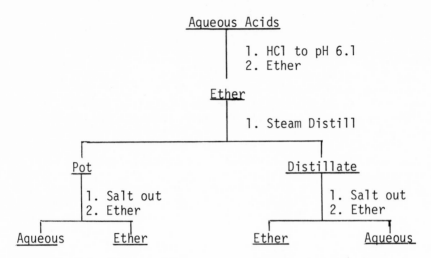

Figure 2.   Scheme 2 - Fractionation of acids

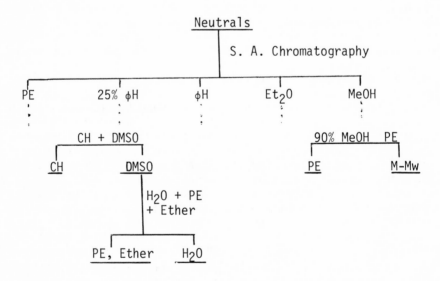

Figure 3.   Scheme 2 - Fractionation of neutrals

linoleic, and linolenic acids as their methyl esters. Also in the same fraction she has identified hydroquinone and catechol by GLC.

Since activity was shown to be present in all three of the neutral fractions from scheme 1, it was decided to try silicic acid (S.A.) column chromatography in order to concentrate the major tumor promoting activity into one fraction or at least find out a little more about the type of compounds causing this activity. As can be seen in Figure 3, the first three fractions from the silicic acid column were partitioned between cyclohexane (CH) and dimethyl sulfoxide (DMSO). Water was then added to the DMSO fraction and this was extracted with successive portions of petroleum ether (PE) and ether (Et$_2$O) which were combined. In using this scheme it was found that more than 97% of the benzo(a)pyrene (BaP) content of smoke was found in one fraction, ØH-PE-DMSO, with small amounts occurring in each of the adjacent DMSO solutions. By this method, it is possible to obtain an enriched fraction of BaP from smoke condensate containing 105 ppm BaP rather than the 1 ppm usually encountered. In additional studies, the ØH-PE-DMSO fraction was further separated on cross-linked polystyrene by lipophilic gel filtration using acetone as developing solvent. The elution curves showed that the polynuclear hydrocarbons could be separated easily from the bulk of the sample weight. Based on the elution pattern obtained, concentrations of at least 10,000 ppm BaP can be obtained by collecting just the fraction containing BaP. This fraction was screened and found to contain principally aromatic hydrocarbons along with the pesticide degradation product TDEE [1-chloro-2,2-bis-(4'-chlorophenyl)ethylene], N-phenyl-4-isopropyl-amine, 9,9-dimethylacridan, and diphenylamine (8).

In the subsequent DMSO fraction, i.e., ØH-DMSO, R. L. Miller has identified 2-methyl, 2,5-dimethyl, 2,3-dimethyl, 2,6-dimethyl, and 2-, 3-, and 4-ethyl benzonitrile along with several alkyl-phenols (9). The occurrence of alkylphenols in the neutral frac-tion may seem unexpected. Closely related compounds have been reported as constituents of the weakly acidic fraction of CSC, such as 2-ethylphenol and 2,6-dimethylphenol. In phenols, the ioniza-tion of the hydroxyl group is influenced usually by inductive, resonance, and steric effects of substituents. Alkyl substitution in the 2- and 6-positions may reduce the acidity or produce other effects so that the phenol is insoluble in aqueous alkali. To determine whether the bulk of the isolated alkylphenols would be expected in the neutral or weakly acidic fractions of CSC, parti-tion coefficients of the isolated compounds in ether-aqueous sodium hydroxide were determined. The results indicated that most of the 2-butylphenol, 2-isobutylphenol, and possibly 2-ethyl-5,6-dimethylphenyl should be in the neutral fraction but substantial amounts of the 2-ethyl-6-methylphenol might be in the weakly acidic fraction depending on the number of successive partition steps employed.

In analyzing the results of the biological testing of fractions from scheme 2 it was found that severe losses in activity were observed when all eluates were combined to obtain a reconstituted neutral fraction. In addition to other factors, it appeared that such losses may have been due at least in part to a failure to elute the columns completely. Even after elution with methanol, all columns showed a tan color indicating that some material remained thereon. To study this problem further, activated and non-activated silicic acid and florisil were employed to separate the neutrals of CSC. Biological data on the strongly adsorbed material indicated that the major loss in activity on recombining chromatographic eluates probably does not arise from this source. In order to avoid this problem it was decided to develop a scheme of separation which eliminated the use of silicic acid column chromatography. We returned to partitioning of the neutrals between 80% methanol:water and cyclohexane as in scheme 1. The cyclohexane fraction was then separated by counter current distribution using a 200 tube instrument. After two hundred partitions between cyclohexane and nitromethane, the tubes were divided into five fractions for bioassay analysis. It was found that the BaP content was in tubes 46 - 140 (this was done by the addition of carbon 14 labeled BaP) while the major portion of the weight was found at either end of the distribution.

By this time we had received the results of the fractionation in scheme 2 and had found that the M-Mw fraction as well as the ØH-PE-DMSO fraction contained a significant amount of activity. This dictated going back to silicic acid chromatography and working further on the M-Mw fraction about which very little is known. In order to do this scheme 4 was developed. The silicic acid column was eluted in the usual manner and M-Mw fraction obtained. This was subjected to lipophilic gel filtration using benzene as the eluting solvent. The gel was a polystyrene gel with a molecular exclusion limit of 2700. A series of compounds of known molecular weights were run on the column for calibration purposes. The major portion of the weight (34%) appears to lay in a molecular weight range between 800 and 450. Molecular weights of these fractions were confirmed by vapor pressure osmometry. Preliminary investigations of these fractions indicate a large number of components of complex structure differing only slightly (length of side chain) and therefore making isolation extremely difficult. The i.r. spectra of the fraction molecular weight 450 to 800 are rather nondescript indicating the presence of an OH group and possibly an ester linkage.

REFERENCES

1. R. L. Stedman, R. C. Benedict, M. Dymicky and D. B. Bailey.
    Beitr. Tabakforsch. 5, 97-103 (1969).
2. Ernest L. Wynder and Dietrich Hoffmann, "Tobacco and Tobacco
    Smoke" Academic Press, New York (1967).
3. I. Schmeltz, C. J. Dooley, R. L. Stedman and W. J. Chamberlain.
    Phytochem. 6, 33-38 (1967).
4. R. L. Miller, L. Lakritz, C. J. Dooley and R. L. Stedman.
    Tob. Sci. 11, 35-36 (1967).
5. A. P. Swain, J. E. Cooper, and R. L. Stedman. Cancer Research
    29, 579 (1969).
6. Fred G. Bock, A. P. Swain and R. L. Stedman. Cancer Research
    29, 584 (1969).
7. R. L. Stedman, R. L. Miller, L. Lakritz and W. J. Chamberlain.
    Chem. and Ind. 394 (1968).
8. R. L. Miller, W. J. Chamberlain and R. L. Stedman. Tob. Sci.
    13, 21 (1969).
9. R. L. Miller, and R. L. Stedman, Phytochem. 10, 1135 (1971).

METHODS FOR BIOASSAYS OF TOBACCO SMOKE

F. G. Bock

Roswell Park Memorial Institute, New York State

Department of Health, Buffalo, New York 14203

Cigarette smoking has been associated with a number of diseases in man. In most cases this relationship has required a long period of exposure and the incidence of disease is dependent upon the number of cigarettes consumed. Attempts to interrupt a sequence between smoking and disease have presented us with a series of very complex problems, most of which have not yet been solved. First of all it is unlikely that any single constituent of cigarette smoke could account for all of the conditions which develop in a group of heavy cigarette smokers. Many separate assay procedures will be required to determine effects of cigarette design on the delivery of potentially hazardous agents in the smoke, and these procedures are not available today. I think it is fair to say that although no fully satisfying laboratory model of the various diseases is available, the models for carcinogenesis offer promise that something positive can be achieved in this area.

The greatest body of experimental data related to causation of chronic disease in man probably consists of the studies of those compounds which cause cancer on contact with animal tissue. The belief that cancer could result from contact with environmental agents was first expressed in 1775 and was first confirmed experimentally in 1915 (1,2). Since 1915 there have been thousands of papers devoted to the study of this phenomenon. It is not surprising, therefore, that the chief experimental investigations of the relationships between cigarette smoke and chronic disease have concentrated on the activity of cigarette smoke condensate as a contact carcinogen. It is almost certain that the carcinogenic activity of cigarette smoke condensate will have been successfully modified by 1980. Indeed, a degree of desirable change may have

107

occured already (3). We can hope that these changes will be
followed by a reduction in the tumor incidence — particularly
the lung tumor incidence — among cigarette smokers. If so, success will be due in large part to the development of useful bioassay techniques over the last 50 years.

PROBLEMS FOR BIOASSAYS

Over the years we have recognized a number of problems in
design and interpretation of bioassays of potential carcinogens.
Some of these problems include (1) metabolic activation of precursor compounds, (2) species and sex differences, (3) co-
carcinogens, (4) thresholds, and (5) other dose-response effects.

Metabolic Activation of "Carcinogens"

We have learned to recognize that before they can produce
tumors, certain compounds must be metabolized in organs remote
from their site of first contact with the body. Although these
substances may properly be considered carcinogens, they are, in
reality, precursors of the ultimate carcinogens which are produced from them. With these agents, the route of administration,
the species and sex of target animal, and the target organ itself
are often critical in experimental design. To test for environmental carcinogens of this sort, it is important to duplicate as
nearly as possible the pattern of exposure to which man is subjected. Other compounds are either carcinogenic in themselves,
or modifed into the ultimate carcinogenic form by virtually all
cells of the body. Carcinogens of this type appear capable of
producing cancer on direct contact. It is very likely that lung
cancer in smokers is a response to such "contact carcinogens".
These contact carcinogens can be studied in a greater variety of
situations than can carcinogens which require metabolic alteration
in specific organs. Nevertheless, species and sex differences and
differences related to route of administration must be considered
in interpretation of experimental results. The cells which are
most likely to develop tumors are those which are exposed to the
highest concentrations of the carcinogen — generally cells at the
point of first contact with the carcinogen. Skin painting or subcutaneous injection are the classic ways to study these agents
primarily because of expense, convenience, and the availability
of a large background literature.

Species and Sex Differences

Either anatomic or metabolic differences can account for
variations in species and sex responses to carcinogens. For ex-

ample, the mouse and the rabbit are particularly sensitive to skin painting by hydrocarbon carcinogens, whereas the rat and the guinea pig are relatively insensitive. Some of this difference in sensitivity is due to the much greater penetration of hydrocarbons into the skin of rabbits and mice as compared to those of the other rodents. Patterns of hair growth may also cause variations in response to carcinogens. In the mouse, rat, hamster, and rabbit, hair grows in waves and thus adjacent folicles are in the same stage of hair growth cycle. In other species such as man, guinea pigs, and the cat, hair growth occurs in a mosaic pattern such that adjacent areas of skin may have folicles in various states of the hair growth cycle. We know that the hair growth cycle affects carcinogenesis profoundly in the absence of drastic changes in the penetration or persistence of carcinogens in the skin. Both the chemical composition and thickness of the skin are directly affected by the hair growth cycle and may account for these effects. Species dependent metabolic factors undoubtedly affect subcutaneous testing as well as skin painting studies. Rats appear to be particularly sensitive to subcutaneous injection of many agents — so sensitive, in fact, that the "negative" vehicle control animals generally develop a few tumors which must be considered in evaluating weak carcinogenic stimuli.

Species differences, of course, imply that laboratory data can imply but cannot certify that an agent is either carcinogenic or non-carcinogenic for man. Laboratory assays, therefore, serve chiefly as guides for the development of less hazardous environments for man; they must be confirmed by observation in man.

Because of toxicity, whole cigarette smoke condensate and some of its fractions cannot be adequately tested by subcutaneous testing. Skin painting is, therefore, used most often today to study these materials. Other routes of administration, more specifically referable to smoking, may be developed later. Mice are most desirable for skin painting studies because they are highly controlled, they are sensitive and because experiments with them are less costly than with other animals. For other routes of administration such as inhalation, mice are less acceptable because of toxicity, anatomic considerations and sensitivity to intercurrent infection.

Cocarcinogenesis

A compilation of different ways in which agents can combine to produce tumors was recently published by Berenblum (4). Whether cocarcinogenesis is involved in human disease is as yet conjectural. It seems probable that combinations of carcinogens having similar structures may act by addition in man just as they

do in experimental animals. We do not know whether other more
specific interactions of carcinogenic stimuli also affect man.
The nature of initiation and promotion have been elucidated by a
number of investigators, among them, Berenblum, Shubik, Saffiotti
and Boutwell (5-8). In mouse skin, a rapid initiation stage is
followed by a protracted stage of promotion. Some compounds, the
complete carcinogens, cause both events to occur. Other compounds
such as urethan are primarily tumor initiators. Still others such
as 12-O-tetradecanoyl-phorbol-13-acetate, the most important
active ingredient of croton oil, are primarily promoters. Theoretically a tumor initiator should not produce tumors in the absence of a promoter and a promoter should not produce tumors in
the absence of an initiator. But practically speaking, I know
of no compound for which this is true. All of the initiators and
all of the promoters we have tested extensively will produce an
occasional skin tumor when these agents are used alone. It is
possible, of course, that the appearance of tumors in mice which
have been treated with either initiators alone or promoters alone
is a manifestation of extremely low levels of environmental initiators and promoters that are not detected in the laboratory.
Background radiation might be such an environmental agent. The
importance of initiation and promotion for the design of assay
systems is that many compounds are capable of producing a large
number of tumors in animals under certain environmental conditions,
but only under these conditions. For example, tumor promoters
are potent "carcinogens" only in mice which have been previously
exposed to an initiator.

Although there is no firm evidence that initiation and promotion play a role in human carcinogenesis, the possibility that
cigarette smoke acts as a tumor promoter has been advanced and
seems to fit certain features of the epidemiology of lung cancer
(9-13). It is prudent, therefore, to examine all fractions of
cigarette smoke condensate (CSC) for both initiating and promoting
activity before considering them to be inactive.

Aqueous extracts of tobacco act as tumor promoters by virtue
of the simultaneous presence of two agents (14). The nature of
their possible interactions between themselves and with the test
animal are not known today. It may thus be necessary some day to
consider types of cocarcinogenesis other than initiation-promotion,
when studying CSC fractions.

Threshold

From a theoretical viewpoint it is impossible to establish
that a given stimulus is incapable of producing tumors. Rather,
one is limited to the demonstration that under certain experi-

mental situations, the stimulus produced no tumors in the limited population at risk. It is very difficult, therefore, to demonstrate that a threshold concentration exists below which a suspected carcinogen cannot produce tumors. Nevertheless, practical limits exist such that the numbers of tumors to be expected in any given situation would be too small to be detected even in large populations. In the case of the potent hydrocarbon carcinogens such as 7,12-dimethylbenz[a]anthracene (DMBA) or benzo[a]-pyrene (BP) such a limit may be very close to zero. Even in relatively unpolluted environments, man may be exposed to levels of these agents above those required to produce occasional malignant transformations. With these agents, any increment of the carcinogenic compound into the environment might be considered hazardous for man albeit with very small risk. In contrast, there is evidence that thresholds of activity exist for tumor promotion by croton oil and its derivatives. The effectiveness of croton oil as a tumor promoter depends on the amount of the agent in each application, and on the frequency of application (8). If this pattern holds true for other tumor promoters, it will be essential to consider the patterns of exposure in man very carefully before one may assume that a particular exposure to a tumor promoter might increase the tumor incidence in the population.

Accommodation of possible thresholds is important in the design of animal tests to detect potential human carcinogenic stimuli. Using equivalent total doses of croton oil, Boutwell found that one application every two weeks was the minimum frequency required for manifestation of promoting activity and that weekly applications were more effective than biweekly applications. Croton oil is a very potent tumor promoter. One should ask, therefore, whether much more frequent exposures are not required for promoters of lesser intensity. Phorbol-12,13-dibenzoate, a weak promoter apparently requires four applications a week to produce tumors in DMBA treated mice (15). Oleic acid is inactive when applied weekly, but is a good promoter when applied 6 times a week (16). Likewise, Tween 60 is a good promoter when applied daily, but not when applied twice a week (17). Historically, many skin carcinogens have been administered to mice three times a week. I have tried to find the rationale behind this schedule without success. (Perhaps the first experiments were conducted by a professor who taught classes three days a week and worked in his laboratory on the other three.) In any event, there is no experimental evidence that a three times a week schedule is optimal for all carcinogens and some evidence suggesting a more frequent treatment is better. Until we have much more experience with a variety of tumor promoters, it is important that high frequency of application be employed. In my laboratory, we generally search for promoters using five applications of the test solution weekly — not because we believe this is best, but because we work on a five day schedule. When more is known, we must arrange our time to fit the experiment rather than the reverse. For tests of CSC or other nicotine containing test materials, frequent application has a second benefit. Nicotine is rapidly absorbed and rapidly detoxified. Frequent application permits greater total dosage without acute lethality. For tests of CSC as a complete carcinogen, we customarily use 10 applications per week.

Dose-Response Effects

Generally speaking, carcinogens act as most other toxins in that the incidence of positive response is dependent upon the dose employed. This is not necessarily the case, however, when toxic properties of the carcinogenic stimulus are important. For example, there appears to be a practical limit in positive response to crude cigarette smoke condensate or relatively crude fractions thereof (18). Until clear dose-response curves are developed for each specific carcinogen in its crude environment, quantitative comparisons of two carcinogenic stimuli would require that at least one of them be tested at two or more concentrations such that the response of one is bracketed by the range of response of the other. This requirement is not as critical in assays designed to identify specific carcinogenic agents in natural products. However, failure to consider dose levels can cause a great deal of unnecessary confusion. One such consequence is the terminology that permeates, not only the scientific literature, but more importantly the press transcription of scientific reports. The term strong carcinogen appears to mean something to all of us. However, this semantic trap has caused a great deal of unwarranted excitement about relatively minor carcinogenic stimuli. For example, everyone would agree that BP is a very strong carcinogen. In contrast, a one part per million solution of BP in acetone is such a weak carcinogenic stimulus that it would be only marginally detectable in simple bioassays. Nevertheless, many investigators have apparently felt that the search for a potential carcinogen in an environmental agent is complete as soon as they find a "strong carcinogen" in the system. The carcinogenic stimulus they have uncovered is often of questionable importance. BP in levels which are not significant carcinogenic stimuli has been reported in such diverse sources as lake water, asbestos, and, of course, air and tobacco smoke (3, 19-21). If we would use the term carcinogen to describe a pure compound in the bottle and carcinogenic stimulus to describe a mixture containing the carcinogen, most of these problems would not trouble us. Many "carcinogens" have been found in tobacco smoke but no one of them in concentrations sufficient to be an important carcinogenic stimulus in any known experimental system. Perhaps a number of individually insignificant carcinogenic stimuli

combine in tobacco smoke so that the mixture is an important carcinogenic stimulus. Evidence that the hydrocarbon carcinogens in
tobacco are important stimuli as a group has been presented (22),
but the hypothesis that every minor carcinogenic stimulus in
tobacco tar contributes its bit to the activity of the whole has
not been confirmed nor is it likely, in my opinion, to prove correct. In general, tests of CSC and its derived fractions should
employ concentrations that are relatively comparable to those found
in the crude material. In this way, the relative importance of
the various carcinogenic stimuli may be determined.

REQUIREMENTS FOR BIOASSAYS

Given the problems associated in general with bioassay of carcinogens, what are the specific requirements that must be met for
bioassays of cigarette smoke? The answer depends on whether the
data will be used to compare different smoking products which may
have different carcinogenic potencies or whether they will be used
to identify specific contributors to the carcinogenic stimulus of
CSC. For empirical evaluation of different formulations, a test
should satisfy the following criteria in order of importance: (a)
highly reliable dose-response effects, (b) both intra- and interlaboratory reproducibility, (c) limited deviation from the human
response and (d) economic feasibility of an adequate number of
comparisons. On the other hand, if one wishes to identify specific
active agents, he will require a test which most fully satisfies
the following criteria: (a) high level of sensitivity, (b) moder-
ately reliable dose-response effects, (c) intralaboratory reproducibility and (d) related to the human response that motivates
the study.

Empirical Comparison of Different Formulations

If one is interested in studying different formulations of
smoking products, a highly reliable dose-response pattern for the
bioassay is a most important criteria. In the mouse skin system
satisfactory dose-response relationships for pure carcinogens and
for CSC have been described in several laboratories (23). Groups
of 100-200 mice will generally be required to disclose reductions
of about one third of the activity of "standard" CSC. Dose-
response studies for other systems designed to disclose contact
carcinogens have not been so well worked out.

It is most desirable that the comparisons be conducted in
different laboratories with different batches of smoking product
but using identical methodology. This would provide a basis for
estimation of intralaboratory and interlaboratory variation in the

studies. Solutions of pure carcinogens give fairly reproducible
results within and among laboratories when identical assays are
conducted. However, tobacco smoke condensates are much more complex and their nature depends on details of preparation. No adequate measure of laboratory variability with mouse skin painting
of tobacco smoke condensates is available because methods differ
among laboratories and because standard reference cigarettes with
constant composition have been available only in the last few
years. Most comparisons of cigarette smoke condensate presently
conducted in the U. S. include Kentucky Reference cigarettes to
permit future determination of these variations.

    Deviation of the test system from human response cannot be
measured until the results of the tests are applied to man. However, it seems self evident that inhalation systems ought to give
the least problems in this respect. Although mouse skin is adequate to identify contact carcinogens, it is conceivable that two
agents will exhibit different relative activities when applied to
skin or lung. This can be true even when the chemical nature of
the carcinogens is similar. For example, when acetone solutions
of 3-methylcholanthrene (MC) and dibenz[a]anthracene (DBA) are
applied to mouse skin, MC is much more active (24). However, when
suspensions of these carcinogens are injected intravenously into
mice to produce lung tumors, DBA is the more active (25). Probably these differences result from the greater solubility of MC.
The more soluble MC penetrates into the skin better whereas the
less soluble crystals of DBA are more efficiently retained in the
lung. The differences in surface between skin and bronchus should
also affect exposure of the target cells through solubility considerations. These differences would be critical should a comparison be made between two condensates one of which was low in one
carcinogen, but high in a second. These considerations require
the development of inhalation systems for the purpose of comparing
different types of smoking products. Several models appear promising (26-28). Inhalation systems would provide a second advantage of providing information concerning pulmonary effects of
smoke other than malignant transformation. At the present time,
however, failure to obtain adequate numbers of tumors with presently developed inhalation systems requires that comparisons of
the carcinogenic effects of cigarette smoke condensates must be
limited for practical consideration to mouse skin assay systems.

    Ordinarily it would seem callous to select methods which are
less than ideal in attempts to solve problems of human morbidity
and mortality. However, the costs of carcinogenesis bioassays of
cigarette smoke condensate are so very great that the size of the
research effort is limited today by availability of facilities
rather than by the availability of ideas. It costs about $80,000.
to compare two cigarettes in a mouse skin system. Although pre-

liminary indications will be available in a year or so, the experiments may drag on for a year and a half or more. These con-
straints have severely limited progress in development of less
hazardous cigarettes. If we had to rely upon present inhalation
systems, the costs would be many times greater and the time required to complete experiments would probably be much greater than
that for skin painting tests.

In summary, mouse skin painting is the single system that
most adequately satisfies the criteria for comparison of different
types of cigarettes. It seems likely that skin tests will be the
chief bioassay system for a decade or so after which they may be
supplanted by as yet undeveloped inhalation methods.

It should be wise to consider one more problem that may develop as we compare cigarettes with different modifications. The
question arises if one is dealing with a multicomponent system
containing both complete carcinogens and tumor promoters. What
will happen if only one component is reduced in quantity? At the
present time there is nobody of experience with reference compounds that would permit us to make any solid predictions. I'm
led to ask this question particularly because we have occasionally
obtained results of cigarette comparisons which suggest that only
one component of a multicomponent system has been reduced. For
example, we have seen experimental cigarettes which over the early
part of the experiment provide substantially fewer tumors than the
control cigarettes. However, late in the experiment the development of tumors appears to be identical between the experimental
and control groups. We speculate that the experimental cigarettes
deliver fewer tumor promoters but the same amount of complete carcinogens. Whether this is true or not cannot be established until
the agents have been identified and until model systems containing
the agents have been submitted to test.

Identification of Specific Active Agents

Subcutaneous injection is a very efficient procedure for producing tumors in experimental animals. With the hydrocarbon carcinogens, microgram quantities are adequate to produce a positive
response. Two considerations, however, argue against its use for
studying cigarette smoke condensate and its fractions. First of
all, crude condensate is so toxic in this system that the activity
of fractions cannot be compared with the crude material from which
they were obtained. Secondly, the method has yielded positive
results when many substances which are clearly not potent carcinogens are tested under unusual conditions. For example, hypertonic
fructose, and glucose solutions produce tumors when injected repeatedly into the same subcutaneous site of mice and rats. This

gross deviation from human experience has tended to discredit the use of subcutaneous testing for routine assays of smoke condensate fractions.

Mouse skin offers a less sensitive site for carcinogenic study. Using total doses of 45 grams of CSC per mouse, we can obtain tumors in about half of the animals over a 75 week period of observation. It would require about 10 cartons of cigarettes to provide this much condensate. To account for losses and separation of important carcinogenic stimuli in tests of fractions, we would have to apply the material from 45, 90, or 180 grams of CSC per mouse. The test is obviously not very sensitive.

An alternative to skin painting procedures that we have employed — successfully, I think — is to examine the various fractions for tumor promoting activity only. This method is based on the assumption that any complete carcinogens which provide an important carcinogenic stimulus in smoke condensate will be more easily detected in animals that have been pretreated with an initiating stimulus consisting of a single application of 125 μgm of DMBA. The promoting type assay has an added advantage because it will also detect fractions which are effective tumor promoting stimuli although they may not be complete carcinogenic stimuli. If cigarette smoke does, indeed, act as a tumor promoter in man, identification of the promoters in smoke condensate is of immediate importance. Even if promoters do not affect man, they must be identified in order that we can evaluate the results of mouse skin tests comparing various cigarettes. Using promoting assay we are able to reduce the amount of condensate required by a factor of 6 without sacrifice of sensitivity. Furthermore, we reduce the duration of the experiment by about 1/3 — a very important advantage when repeated series of tests are required during stepwise fractionation. The positive fractions that are ultimately identified may be either tumor promoters or complete carcinogens, and it will be necessary, therefore, to test them in the absence of DMBA pretreatment in order to distinguish between these alternatives. If one were dealing with a simple fractionation, it would be faster and more efficient to test the samples as complete carcinogens and as promoters concurrently. However, for the fractionation of smoke condensate, a large series of sequential fraction steps are required. It is faster and much more economical to first identify the active fractions, obtain them in a relatively pure form and then finally characterize them according to the nature of their activity.

With promotion type assays, we have found that dose-response effects are not always what we would desire. This may be due to a unique property of tumor promotion or it may be due to the fact that the initiating stimulus is not uniformly applied to all of the animals. Regardless of the source of the problem, comparisons among groups are less valid in a promotion type assay than in assays of the complete carcinogenic activity. Nevertheless, the economy and broader scope of promotion assays strongly support their use.

The requirements that an assay to detect specific carcinogenic stimuli in CSC be related to the development of tumors in man has limited most bioassays to mouse skin painting either in the complete carcinogenesis or in the tumor promoting system. A number of simpler and more economical tests should be examined to see if they might not provide information nearly as valuable with a great deal less effort. Just as promoting tests offer an opportunity to isolate active materials that can later be put to a complete carcinogenesis assay, in vitro or simple in vivo systems might identify agents that could later be tested for either complete carcinogenic activity or tumor promoting activity. Systems that offer promise include subcutaneous injection of fractions which are not toxic, cell culture techniques resulting in transformation or other aberrant growth, chemical potentiation of viral effects in either cell culture or whole animals, and simple chemical reactivity. It ought to be possible to compare the effects of various fractions that have been characterised on mouse skin with these various potential test systems. If a simplier test system gives a high correlation with the mouse skin system when applied to a large variety of fractions, the simple system should be employed before the more elaborate and expensive skin assay.

When the tumor promoting assays are applied to the fractions of CSC that have been described by Chamberlain (29), at least five clearly distinguishable fractions that are significant tumor promoting stimuli can be detected. One of these is the nonvolatile weak acid fraction. The whole weak acid fraction is only a marginal complete carcinogenic stimulus when applied in concentrations comparable to those in CSC (30). It is, therefore, properly considered a typical tumor promoter. A second fraction contains polar neutral materials. This fraction does not appear to have tumor initiating activity and, accordingly, may also be considered a tumor promoter (31). One fraction which contains nearly all of the BP of CSC is an active tumor initiator as well as an active tumor promoter. It represents the most important complete carcinogenic stimulus in smoke condensate. The other two active tumor promoting fractions are eluted from silicic acid in adjoining fractions to that containing BP. When tested for initiating activity, these fractions were inactive; however, it is possible that the lack of activity was due to the fact that they were not tested in adequate concentrations. They represent relatively minor promoting stimuli in CSC and could conceivably be relatively minor complete carcinogenic stimuli.

The true nature of the compounds responsible for these activities cannot be elucidated without further study. We anticipate
that the molecular weight range of the polar neutral fraction will
be disclosed by a current experiment which ought to be completed
in the next several months (29). We, likewise, hope that the
nature of the effective tumor promoters in the nonvolatile weak
acid fraction will also be clarified. The identity of the important carcinogenic stimuli that occur in the fractions that be-
have similar to BP may require much more intensive investigation.
Hoffmann and Wynder have reported that a similar fraction provides
a large number of subfractions many of which have tumor initiating
activity and presumably complete carcinogenic activity (22).

In summary, the use of mouse skin tests as complete carcinogenic assays of CSC from various types of cigarettes appears to be
straight forward and meets most of the criteria required for the
assay system. The test system differs somewhat from the human
situation and we can hope that in the years to come alternative
procedures more related to human exposure can be adequately developed. The use of the mouse skin system particularly in a promotion type assay offers a more efficient test for active agents
to further their isolation and ultimate identification. The method
is not as sensitive as we desire; other systems — particularly
in vitro systems — should be developed.

It would be gratifying if I could stop at this point. However, I feel I must introduce the problem of our ignorance concerning the effects of co-carcinogens acting in concert — even in
a system so thoroughly studied as the mouse skin. We know from
preliminary studies that CSC is much less active as a carcinogen
than a simple mixture containing an equivalent amount of initiator
and promoter. Does this indicate that combinations have some
unique interaction other than those we customarily expect with
initiation and promotion? Does it mean that crude CSC contains
agents which interfere with complete carcinogenic activity but not
initiation or promotion effects? What would happen if we made the
system even more complex by including two or three different types
of promoters? These are all problems that have to be answered
before we can fully understand the test systems we employ.

REFERENCES

1.   Pott, P. Chirurgical Observations Relative to the Cataract,
      the Polypus of the Nose, the Cancer of the Scrotum, the
      Different Kinds of Ruptures and the Mortification of the Toes
      and Feet. London, Hawes, Clarke, & Collins, 1775.

2.   Yamagiwa, K. and Ichikawa, K. Experimental Study of the
      Pathogenesis of Carcinoma. J. Cancer Res. 3:1-21, 1918.

3.   Wynder, E. L. and Hoffmann D. Tobacco and Tobacco Smoke.
     Studies in Experimental Carcinogenesis. New York, Academic
     Press, Inc., 1967.

4.   Berenblum, I. A Re-evaluation of the Concept of Cocarcino-
     genesis. Progr. Exper. Tumor Res. 11:21-30, 1969.

5.   Berenblum, I. The Mechanism of Carcinogenesis: A Study of
     the Significance of Cocarcinogenic Action and Related
     Phenomena. Cancer Res. 1:807-814, 1941.

6.   Berenblum, I. and Shubik, P. A New Quantitative Approach to
     the Study of Stages of Chemical Carcinogenesis in the Mouse's
     Skin. Brit. J. Cancer 1:383-391, 1947.

7.   Saffiotti, U. and Shubik, P. The Effects of Low Concentrations
     of Carcinogen in Epidermal Carcinogenesis. A Comparison with
     Promoting Agents. J. Nat. Cancer Inst. 16:961-969, 1956.

8.   Boutwell, R. K. Some Biological Aspects of Skin Carcino-
     genesis. Progr. exp. Tumor Res. 4:207-250, 1964.

9.   Eastcott, D. F. The Epidemiology of Lung Cancer in New
     Zealand. Lancet 1:37-39, 1956.

10. Dean, G. Lung Cancer Among White South Africans. Brit. Med.
     J. 2:852-859, 1959.

11. Doll, R. Interpretations of Epidemiologic Data. Cancer
     Res. 23:1613-1623, 1963.

12. Selikoff, I. J., Hammond, E. C., and Churg, J. Asbestos
     Exposure, Smoking, and Neoplasia. J. Am. Med. Assoc.
     204:106-112, 1968.

13. Bair, W. J. Inhalation of Radionuclides and Carcinogenesis.
     In: Inhalation Carcinogenesis, ed. M. G. Hanna, Jr.,
     P. Nettesheim, and J. R. Gilbert, pp. 77-101, USAEC Div. of
     Tech. Inform., Oak Ridge, Tenn., 1970.

14. Bock, F. G. The Nature of Tumor-promoting Agents in Tobacco
     Products. Cancer Res. 28:2363-2368, 1968.

15. Baird, W. M. and Boutwell R. K. Tumor-promoting Activity of
     Phorbol and Four Diesters of Phorbol in Mouse Skin. Cancer
     Res. 31:1074-1079, 1971.

16. Holsti, P. Tumor Promoting Effects of Some Long Chain Fatty
     Acids in Experimental Skin Carcinogenesis in the Mouse. Acta
     Path. microbiol. Scand. 46:51-58, 1959.

17. Setälä, K., Setälä, H., Merenmies, L. and Holsti P.
     Untersuchunger über die tumorauslosende ("tumour promoting")
     Wirkung einiger nichtionisierbaren oberflächenaktiven
     Substanzen bei Maus und Kaninchen. Z. Krebsforsch. 61:
     534-547, 1957.

18. Muñoz, N., Correa, P., and Bock, F. G. Comparative Carcino-
     genic Effect of Tow Types of Tobacco. Cancer 21:376-389,
     1968.

19. Borneff, J., and Fischer, R. Kanzerogene Substanzen in
     Wasser und Boden XI. Polyzyklische, aromatische Kohlenwas-
     serstoffe in Walderde. Arch. Hyg. Bacteriol. 146:430-437,
     1962.

20. Harington, J. S. Chemical Studies of Asbestos. Ann. N. Y.
     Acad. Sci. 132:31-47, 1965.

21. Sawicki, E., Hauser, T. R., Elbert, W., Fox, F. T., and
     Meeker, J. E. Polynuclear Aromatic Hydrocarbon Composition
     of the Atmosphere in Seven Large American Cities. J. Am.
     Ind. Hyg. Assoc. 23:137-143, 1962.

22. Hoffmann, D., and Wynder, E. L. A Study of Tobacco Carcino-
     genesis XI. Tumor Initiators, Tumor Accelerators, and Tumor
     Promoting Activity of Condensate Fractions. Cancer 27:
     848-864, 1971.

23. Bock, F. G. Dose Response: Experimental Carcinogenesis.
     Nat. Cancer Inst. Mono. 28:57-63, 1968.

24. Bock, F. G. Early Effects of Hydrocarbons on Mammalian Skin.
     Progr. exp. Tumor Res. 4:126-168, 1964.

25. Shimkin, M. B. and Lorenz, E. Factors Influencing the
     Induction of Pulmonary Tumors in Strain A Mice by Carcino-
ge   genic Hydrocarbons. J. Nat. Cancer Inst. 2:499-510, 1942.

26. Rockey, E. E. and Speer, F. D. The Ill Effects of Cigarette
     Smoking in Dogs. Intern. Surg. 46:520-530, 1966.

27. Dontenwill, W. Experimental Investigations on the Effect of
     Cigarette Smoke Inhalation on Small Laboratory Animals. In:
     Inhalation Carcinogenesis, ed. M. G. Hanna, Jr., P. Nettesheim,
     and J. R. Gilbert, pp. 389-412, USAEC Div. of Tech. Inform.,
     Oak Ridge, Tenn., 1970.

28. Auerbach, O., Hammond, E. C., Kirman, D. and Garfinkel, L.
     Effects of Cigarette Smoking on Dogs. II. Pulmonary Neo-
     plasms. Arch. of Env. Health 21:754-768, 1970.

29. Chamberlain, W. J. and Stedman, R. L. Fractionation of
     Tobacco Smoke Condensate for Chemical Composition Studies.
     This monograph.

30. Bock, F. G., Swain, A. P., and Stedman, R. L. Composition
     Studies on Tobacco. XLIV. Tumor-Promoting Activity of Sub-
     fractions of the Weak Acid Fraction of Cigarette Smoke
     Condensate. J. Nat. Cancer Inst. 47:429-436, 1971.

31. Bock, F. G., Swain, A. P., and Stedman, R. L. Unpublished.

CHEMICAL COMPOSITION AND TUMORIGENICITY OF TOBACCO SMOKE*

Dietrich Hoffmann and Ernest L. Wynder

Division of Environmental Carcinogenesis

American Health Foundation, New York, N.Y.

Numerous biological studies have established the carcinogenicity of tobacco smoke. (1,2) Laboratory experiments identified three types of components which contribute to the overall carcinogenicity of the smoke. These are tumor initiators, tumor accelerators, and tumor promoters. (3)

In order to identify the tumorigenic agents in inhalants one separates the gas phase from the particulate phase. In respect to tobacco smoke this separation is often accomplished by the use of Cambridge filters. Although this separation is somewhat arbitrary and causes some artifact formation, it has proved to be most helpful for the identification of tumorigenic agents in tobacco smoke.

GAS PHASE

The gas phase of cigarette smoke does not induce carcinoma of the respiratory tract in the experimental animal. (4,5) However, this lack of activity does not prove the absence of tumorigenic agents in the gas phase. It rather indicates that the concentration of all volatile carcinogens, if present, is below the threshold level of a carcinogen.

A priori, two approaches appear promising for the identification of volatile carcinogens. One is the systematic bioassay of gas phase fractions. Model studies, however, have shown that fractionations of the gas phase can lead to a number of artifacts, such as the reaction of volatile agents to form new non-volatile components, (6) and to the in vitro nitrosation of secondary amines.

*Chemical Studies on Tobacco Smoke. XV.

(1,7)   The second and more promising approach is the search for
volatile carcinogens which theoretically could be present in tobacco smoke.   This group of volatile carcinogens includes arsine,
nickel carbonyl, nitro olefins, and N-nitrosamines. (Table I)

Table I

SOME THEORETICALLY POSSIBLE CARCINOGENS IN THE GAS PHASE OF
TOBACCO SMOKE

Carcinogens	Chemical Formula	Reference to Carcinogenicity
Arsine	H_3As	Holland and Acevedo, 1964
Nickel Carbonyl	$Ni(CO)_4$	Sunderman and Donnelly, 1965
Nitro Olefins	$R-C=CH-R'$ $\mid$ NO_2	Deichmann et al, 1965
Volatile Nitrosamines	$\begin{array}{c} R-CH_2 \\ R'-CH_2 \end{array}\!\!>\!N-NO$	Magee and Barnes, 1966 Druckrey et al, 1967

Arsine, Nickel Carbonyl and Nitro Olefins

The burning cone of a cigarette produces a reducing atmosphere
containing about 8 volume percent of hydrogen. (13)   Since tobacco
contains arsenic compounds, (1) arsine may be formed, a possibility
not previously investigated.   One possible reason for the lack of
interest in arsine is the drastic decrease of the amount of arsenic compounds in tobacco over the last decade. (14)

Tobacco, like all plants, contains traces of nickel. (1,15-18)
Although several investigators found nickel in the mainstream smoke
of cigarettes, (1,15,17,19) it appears unlikely that it is present
as nickel carbonyl.   A neutron-activation analysis of the gas phase
is required to definitely verify the presence or absence of nickel
carbonyl in tobacco smoke.

The mainstream smoke of cigarettes contains nitrogen oxides,
(1,15) a large spectrum of volatile unsaturated hydrocarbons,(1,15,20)

nitroparaffins, nitrobenzenes, and possibly, nitrophenols. (21-23)
These data suggest that tobacco smoke can be regarded as a possible
environment for the formation of traces of carcinogenic nitro olefins. Again, chemical-analytical studies are indicated.

N-Nitrosamines

The carcinogenic N-nitrosamines have attracted the great
interest of the tobacco scientists. Neurath et al identified N-
nitrosamines in tobacco smoke condensate as early as 1964. (24)
However, this and subsequent studies did not exclude the possibility that the identified nitrosamines were formed as artifacts.
(1,7,15,25-27) Cigarette smoke contains nitrogen oxide (NO) and
volatile secondary amines, but is practically free of nitrogen
dioxide, (1,15,28,29) the reagent which is essential for the formation of nitrosamines. Since the smoke which leaves the burning
cone is free of NO_2, its formation can only occur in a second or
in the fraction of a second during which the smoke travels through
the tobacco column and is diluted with the air which diffuses
through the cigarette paper. The moment the mainstream smoke
leaves the mouth piece, which is connected with the smoking machine,
it is diluted with air. This is known to lead to artificial formation of NO_2 and, consequently, N-nitrosamines. (1,24)

In our analysis of nitrosamines we have excluded artificial
formation of nitrosamines by leading the smoke through a sodium
hydroxide solution and by blowing nitrogen through the collection
vessel immediately after the puff was taken. (Fig. 1) This is
done by using the Borgwaldt 20 cigarette smoker with a rotating
head, by smoking cigarettes only on every second channel, and by
connecting the remaining 10 smoking channels to a nitrogen source.
(3) This arrangement eliminates "aging" of the smoke.

When 300 cigarettes have been smoked, the volatile N-nitrosamines are enriched by distillation and extractions, and are reduced with diborane (B_2H_6). (Fig. 2) The resulting unsymmetrical
hydrazines are condensed with 3,5-dinitrobenzaldehyde and the resulting hydrazones are analyzed by gas chromatography with either
an electron capture detector or a flame ionization detector. With
the aid of mass spectrometry we have identified dimethylnitrosamine,
methylethylnitrosamine, and N-nitrosopiperidine. For the quantitative analysis we use ^{14}C-labelled dimethylnitrosamine as an internal standard. The details and results of this study by Dr. J.
Vais of our laboratory will be presented at the 25th Tobacco
Chemists' Research Conference.

Fig. 1   Smoking machine with collection device for the
          analysis of N̲-nitrosamines in cigarette smoke.

Organic Radicals

Since tobacco smoke is a combustion product of organic matter,
it contains free radicals, (1,15,31-34) but the chemical structure
of these radicals has not been elucidated.   In the only test yet
conducted free radicals showed no carcinogenic activity. (35)   The
lack of experimental data precludes a discussion of the possible
role of organic radicals in tobacco carcinogenesis.

PARTICULATE MATTER

The first large-scale production of epidermoid cancer in experimental animals with tobacco smoke condensate was reported in
1953 (36) and has since been verified in many laboratories. (1,2)
During the past decade literally dozens of studies have been direct-
ed towards the isolation and identification of tumorigenic agents
in the particulate matter of tobacco smoke. (1-3,37-39) All bioassays showed the particulate matter of tobacco smoke to be a com-

ANALYSIS OF VOLATILE N-NITROSAMINES IN CIGARETTE SMOKE

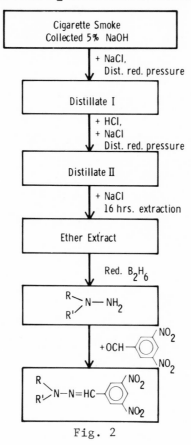

Fig. 2

plete carcinogen, that the activity cannot be explained by a single complete carcinogen or group of carcinogens, that the tobacco "tar" contains tumor initiators and tumor promoters. A significant reduction in the concentration of the tumor initiators or tumor promoters has been shown to lead to a significant reduction in the carcinogenicity of the condensate. (40,41)

Tumor Initiators

Fractionation experiments and bioassays have shown that none of the major fractions are responsible by themselves for the overall carcinogenic activity, or sarcogenicity of tobacco smoke condensate. Only the neutral portion has been shown to induce a significant number of tumors in the experimental animal. (1,3,37,42-44) This observation led to a large-scale fractionation of the neutral portion of cigarette smoke. (3,42) None of the subfractions were

carcinogenic when tested in the concentration representing their
presence in the total "tar". The column chromatography fraction
which eluted after the paraffins contained the polynuclear aromatic
hydrocarbons (PAH) and was carcinogenic when applied in 5 to 10
times the concentration as present in the total condensate. (Fig.3)

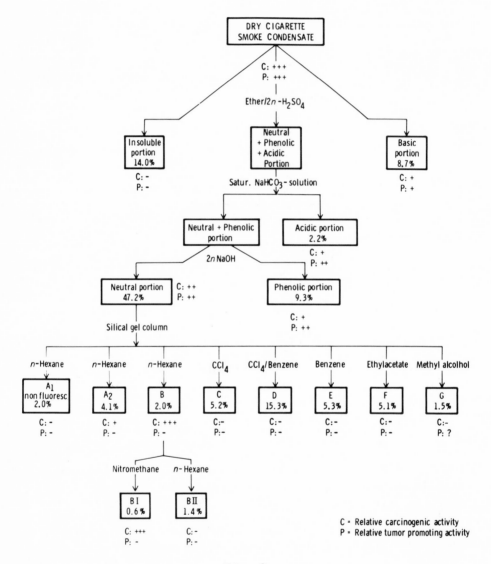

Fig. 3

Recombining all fractions, except the PAH containing subfraction B, led to a reduction of the tumorigenicity of the original "tar" by more than 50%, although fraction B amounted to only 2% of the total "tar". (3,45) In a control experiment, we recombined all fractions including fraction B and observed only a slightly lower carcinogenicity than for the original unfractionated smoke condensate. These results were recently confirmed by Dontenwill and his associates. (37)

The above experimental data suggested that the polynuclear aromatic hydrocarbons (PAH) in tobacco "tar" serve as tumor initiators. To substantiate this we carried out two further experiments. In the first, we reduced the concentration of the polycyclic hydrocarbons in cigarette smoke by 40-60% by adding alkali nitrates to the tobacco, and by 40% in the "tar" from cigarettes produced from tobacco sheets derived from the original tobacco. The tobacco sheets had different burning characteristics. (Figs.4, 5; 46,47) While the total carcinogenicity was reduced by at least

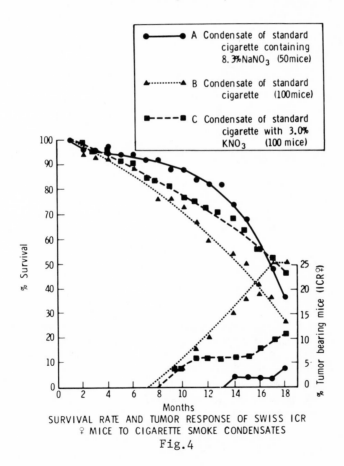

A Condensate of standard cigarette containing 8.3%NaNO$_3$ (50mice)

B Condensate of standard cigarette (100mice)

C Condensate of standard cigarette with 3.0% KNO$_3$ (100 mice)

% Survival

Months

% Tumor bearing mice (ICR♀)

SURVIVAL RATE AND TUMOR RESPONSE OF SWISS ICR ♀ MICE TO CIGARETTE SMOKE CONDENSATES

Fig.4

TUMOR PROMOTING ACTIVITY OF CIGARETTE SMOKE CONDENSATES*

"Tar"	Mice started	After 14 months promotor application			
		Papilloma bearing mice	Total Number papillomas	Carcinoma bearing mice	Survivors
■——■ Standard	60	26	55	9	8
●······● Cigarettes · 5%Cu(NO₃)₂5H₂O	50	24	59	8	14
○——○ Cigarettes · 8.3%NaNO₃	60	21	24	6	24

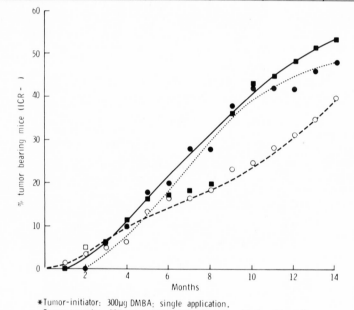

*Tumor-initiator: 300μg DMBA; single application.
Tumor-promoter: 50% smoke condensate; thrice weekly for 12 months.

Fig. 5

50%, the tumor promoting activity of these "tars" was hardly reduced at all. In another experiment we distributed fraction B
between nitromethane and cyclcohexane and obtained a concentrate
BI from the nitromethane layer in which the PAH were further enriched . (Fig. 3) This subfraction was highly active on mouse
skin in concentrations of 5 and 10% and slightly active in a concentration of 2.5%, whereas BII was inactive. (3) BI was highly
active as a tumor initiator in a test with 2.5% croton oil and 50%
tobacco "tar" as tumor promoters.

These experiments supported our working concept that the PAH
are important tumor initiators in tobacco carcinogenesis and are

of decisive importance for the overall carcinogenicity.   In a
study with Gunter Rathkamp, we therefore concentrated the PAH
further for the identification of active agents.   This was done by
gas chromatography and distribution between three pairs of solvents.
(Fig. 6)

     Subfraction BIh amounted to about 0.09% of the total condensate
and contained about 50% of the PAH from the original "tar". (3)
BIh was finally chromatographed on alumina to give 80 subfractions.
(Fig. 7)   There was significant tumor initiator activity in three
groups of fractions, namely BIh 56-66, BIh 67-71, and BIh 72-78.

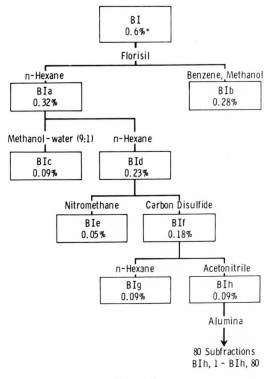

SUBFRACTIONATION OF NEUTRAL FRACTION BI

Fig. 6

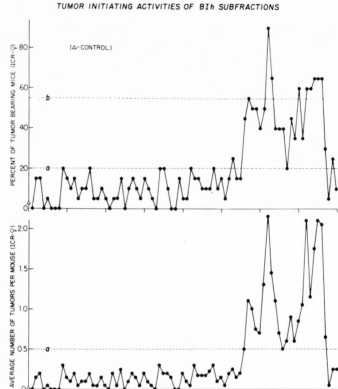

Fig. 7   Each of the fractions 1-80 was applied as tumor initiator in 10 subdoses. a-line: fractions with values
on/or above have significantly higher activity than
the control. (P   0.05)

The latter group (BIH 72-78) contained a large number of known
tobacco carcinogens, e.g. benzo(a)pyrene, and benzofluoranthenes,
as well as a large spectrum of alkylated chrysenes and benzo(a)-
pyrenes. (Table II) Dr. W. Bondinell of our laboratory is synthe-
sizing some of these alkylated polycyclic hydrocarbons to aid in
the identification of the "tar" constituents and they will, subsequently, be tested both as tumor initiators and complete carcinogens.

Table II

MAJOR COMPOUNDS IN TUMOR INITIATING FRACTIONS BIh 71-78

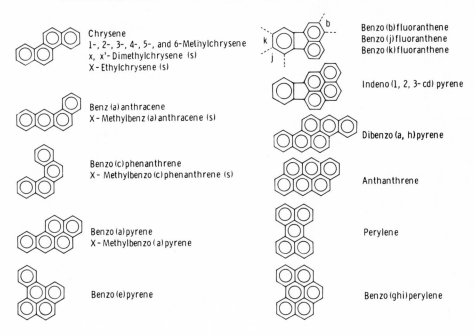

Chrysene
1-, 2-, 3-, 4-, 5-, and 6-Methylchrysene
x, x'-Dimethylchrysene (s)
X-Ethylchrysene (s)

Benz (a) anthracene
X-Methylbenz (a) anthracene (s)

Benzo (c) phenanthrene
X-Methylbenzo (c) phenanthrene (s)

Benzo (a) pyrene
X-Methylbenzo (a) pyrene

Benzo (e) pyrene

Benzo (b) fluoranthene
Benzo (j) fluoranthene
Benzo (k) fluoranthene

Indeno (1, 2, 3-cd) pyrene

Dibenzo (a, h) pyrene

Anthanthrene

Perylene

Benzo (ghi) perylene

Subfractions BIh 56-66 contain chlorinated hydrocarbon insecticides and some of their major pyrolysis products, N-alkyl carbazoles, fluoranthenes, benzofluorenes, cyclopenta(a)phenanthrenes, and alkylated pyrenes. (Table III; 48-50) None of the identified agents were known to be tumorigenic. We have synthesized the five methylfluoranthenes and several dimethylfluoranthenes, and identified several of them in tobacco smoke. Three of them are highly active as carcinogens and tumor initiators. (50) Other agents in BIh 56-66 which have been tested, namely, the insecticides and their pyrolysis products and the N-alkyl carbazoles were, however, inactive as tumor initiators and tumor promoters. The cyclopenta-(a)phenanthrenes amount to only a few percent of these active BIh fractions and even if we assume that they were all active as tumor initiators, we still could not explain the total activity of these fractions.

Tumor Accelerators

This consideration suggested to us that some of these inactive

Table III

MAJOR COMPOUNDS IN TUMOR INITIATING FRACTIONS BIh 55-66

Cl—⬡—CH—⬡—Cl   DDT
         | o, p'- DDT
         CCl₃   DDE (DDT - HCl)

Cl—⬡—CH—⬡—Cl   DDD
         | o, p'- DDD
         CHCl₂   DDM (DDD - HCl)

Trans- 4. 4'- Di-
chlorostilbene

N - Alkylcarbazoles
   N - Methylcarbazole
   N - Ethylcarbazole
   1, N-, 2, N-, 3, N- and
   4, N - Dimethylcarbazole

Fluoranthene
   1-, 2-, 3-, 7-, and
   8 - Methylfluoranthene
   X - Ethylfluoranthene
   x, x' - Dimethylfluoranthene
   Benzo (mno) fluoranthene

Benzofluorenes
   11H - Benzo (a) fluorene
   11H - Benzo (b) fluorene
   11H - Benzo (c) fluorene

17H - Cyclopenta (a) phenanthrenes
   17H - Cyclopenta (a) phenanthrene
   X - Methyl - 17H - Cyclopenta (a)-
       phenanthrene
   X - Ethyl - 17H - Cyclopenta (a) -
       phenanthrene

Pyrenes
   Pyrene
   1-, 3- and 4- Methylpyrene
   x, x' - Dimethylpyrene

X - Phenylindene

agents may accelerate the activity of known carcinogens and tumor initiators. Tumor accelerators are defined as agents, which by themselves are inactive as carcinogens, tumor initiators, and tumor promoters, which, however, can accelerate the activity of carcinogens and/or tumor initiators. (41) We found that trans-4,4'-dichloro-stilbene, a major pyrolysis product of DDT and DDD, and the alkylated carbazoles are active as tumor accelerators. (Figs.8, 9; 3, 41) Although very little is known about tumor accelerators, preliminary data of Dr. P. Chan from our institute suggest that the tumor accelerators inhibit the polycyclic hydrocarbon hydroxylase, thus inhibiting the detoxification of the carcinogenic aromatic hydrocarbons and accelerating their activity. (51)

Tumor Promoters

The first fractionation studies of tobacco smoke condensate showed that the non-carcinogenic acidic portion can significantly increase the tumorigenicity of the neutral portion. (42) In 1958,

Gellhorn reported a significant tumor promoting activity for tobacco
smoke condensate. (52) Since then several laboratories in the
United States and England have confirmed these findings and demonstrated that an important and specific feature of tobacco "tar" is
its tumor promoting activity. (1,2,39,40,53,54) Since Dr. F. Bock
discussed tumor promoting studies yesterday, (55) we will limit our
discussion to recent experiments from our laboratory on the chemical
nature of the tumor promoters.

In 1961 we reported significant tumor promoting activity for
the weakly acidic portion of tobacco smoke condensate, (40,53) a
finding subsequently confirmed by Bock and others. (54,55) We did
not, however, find promoting activity for any of the neutral subfractions. The neutral PAH fraction (fraction BI) initiated with
7,12-dimethylbenz(a)anthracene (DMBA) exhibited the same activity
as fraction BI on tissues which were not initiated with DMBA.(3)
In the case of the most polar neutral fraction it may be that the
elution of the neutral fraction was incomplete, or that the most
polar neutral compounds of tobacco "tar" are oxidized or otherwise
altered on the column, an artifact process well known to occur. (1)

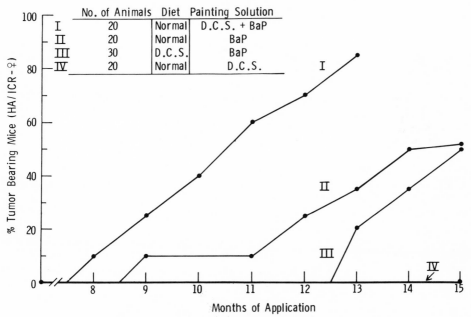

Fig. 8   Tumor accelerating activity of trans-4,4'-
dichlorostilbene (DCS)

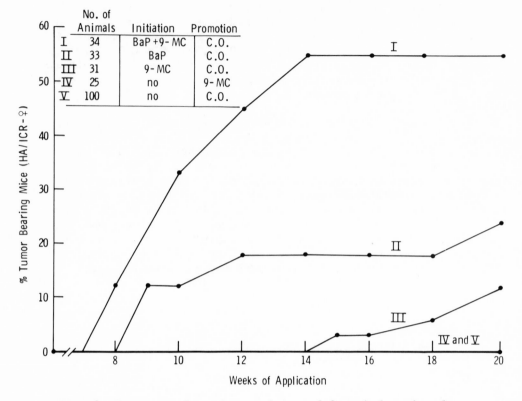

Fig. 9   Tumor accelerating activity of 9-methyl-carbazole

BaP initiation   – 5 μg in 25 μl   10 subdoses of 0.02%
                   solution
9-MC initiation – 0.5% solution
C.O. promotion   – 1.0% solution
9-MC promotion   – 0.5% solution

BaP = benzo(a)pyrene;   9-MC = methylcarbazole;   C.O. = croton oil

A careful fractional distillation of the weakly acidic portion
of tobacco "tar" and, subsequent, testing of the subfractions as
tumor promoters revealed that some activity resides in the fraction
containing the volatile phenols, but that the major promoting activity resides in the non-volatile portion, i.e., the distillation
residue. (Fig. 10; 54,55) About 25% of this fraction consisted
of C_{12} – C_{26} saturated and unsaturated fatty acids. (56) Although

FRACTIONATION OF CIGARETTE SMOKE CONDENSATE

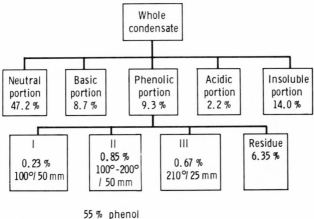

55 % phenol
35 % cresols, dimethylphenols

Fig. 10

several of these fatty acids are known tumor promoters if applied in high concentrations, (57) these acids and the volatile phenols together cannot explain the total tumor promoting activity of the weakly acidic portion.

Recently, Dr. G.Singer from **our** laboratory fractionated the weakly acidic portion and found 3 subfractions, which were strong skin irritants. (Fig. 11; 67) One of the active subfractions contains volatile phenols, while another contains N-alkyl aminophenols, some of which we are presently testing for tumor promoting activity.

It is hoped that this type of study, which is complementary to the work of Dr. F. Bock, will lead to the elucidation of the chemical structure of the tumor promoters. The identification of the promoters should lead to a new approach for the reduction of the carcinogenicity of tobacco smoke, a goal which is foremost in the mind of many tobacco scientists.

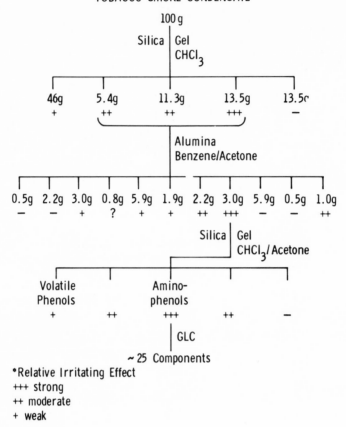

FRACTIONATION OF WEAKLY ACIDIC PORTION OF
TOBACCO SMOKE CONDENSATE

Fig. 11

BLADDER CARCINOGENS

Finally, a few words about the possibility that tobacco smoke
contains bladder carcinogens. A recent report by the Surgeon General of the U.S. Public Health Service states that "epidemiological
studies have demonstrated an association of cigarette smoking with
cancer of the urinary bladder among men". (2) In the experimental
setting, neither passive inhalation of tobacco smoke by mice, rats,
or hamsters, nor painting of mouse skin with tobacco "tar" nor subcutaneous application of "tar" in rats has so far induced bladder
tumors. (1,2) However, a few years ago tumors were found in the
pelvis of the kidneys of a few mice out of a large number of animals
whose skin had been painted with an extract of tobacco "tar". (58)

This rather inconclusive result on the induction of bladder cancer by tobacco products may partially explain why tobacco chemists have so far given little attention to the isolation of bladder carcinogens in tobacco smoke.

Rather detailed studies by three groups of investigators on the presence of known bladder carcinogens led to the isolation of an aminofluorene and α- and β-naphthylamines. (59-61) The classical bladder carcinogen, β-naphthylamine, was found by us in the mainstream smoke of a polpular blended U.S. cigarette in a concentration of only 22 nanograms per cigarette. (61) Although tobacco smoke may contain traces of other carcinogenic aromatic amines, it appears unlikely that these trace amounts can be correlated with the increased risk of cancer of the urinary tract for the cigarette smoker.

A few years ago, Boyland suggested that metabolites of nicotine, nicotine N-oxides, and cotinine are bladder carcinogens. (Fig. 12; 62) However, until these metabolites are proven to be carcinogenic to the bladder of the experimental animal, this concept remains speculative.

METABOLITES OF NICOTINE IN URINE

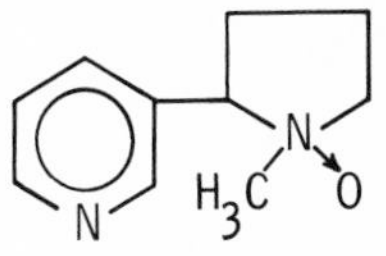

Nicotine N'-oxide

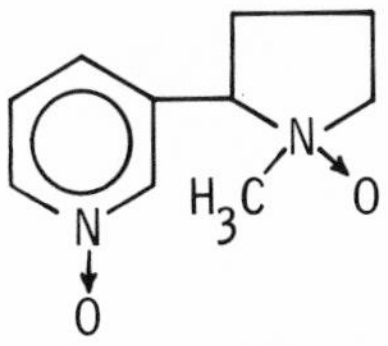

Nicotine N, N'-dioxide

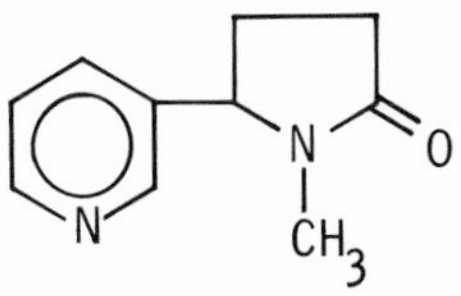

Cotinine

Fig. 12

The more likely mechanism, which may explain the higher risk of bladder cancer for cigarette smokers is the inhibition of the metabolic conversion of hydroxy anthranilic acid derivatives to methylnicotinamide which may result in increased levels of 3-hydroxy anthranilic acid and 3-hydroxykynurenine (Fig. 13; 62) These metabolites of tryptophan are carcinogenic when implanted into mouse bladders. (63, 64)   The urine of heavy cigarette smokers contains an average of 50% more of these metabolites than the urine of non-smokers. (65)

METABOLISM OF TRYPTHOPHAN

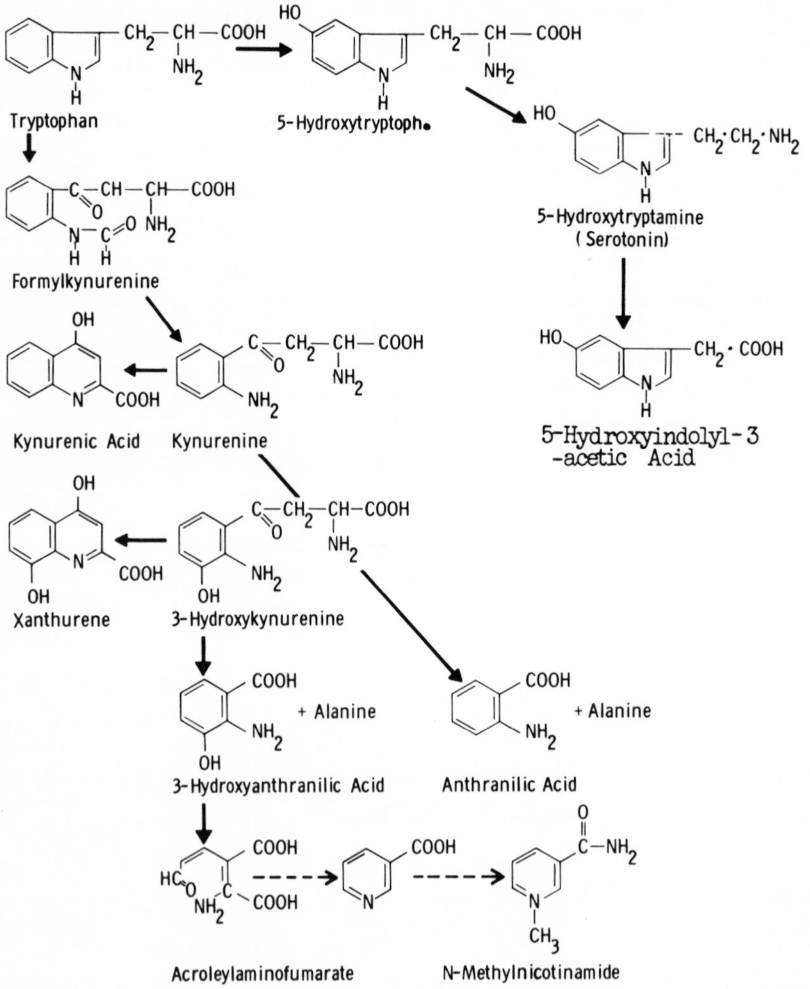

Fig. 13

SUMMARY

Tobacco smoke is carcinogenic and sarcogenic to a variety of animals and tissues. The unaged gas phase separated from tobacco smoke by an inert glass filter was found to be non-carcinogenic. Nevertheless, tobacco smoke is suspected to contain traces of volatile carcinogens such as arsine, nickel carbonyl, nitro olefins and N-nitrosamines. So far, only traces of three carcinogenic N-nitrosamines have been identified in unaged cigarette smoke.

The carcinogenicity of the particulate matter of tobacco smoke is primarily explained by three types of tumorigenic agents: tumor initiators, tumor accelerators, and tumor promoters. The majority of the tumor initiators are polynuclear aromatic hydrocarbons, and, as recently found, a number of alkylated polynuclear aromatic hydrocarbons. A significant inhibition of the pyrosynthesis of these carcinogenic polycyclics leads to a significant reduction of the tumorigenicity of tobacco smoke condensate as does the experimental quantitative reduction of these hydrocarbons in the "tar".

The particulate matter also contains tumor accelerators. These agents have a polarity similar to the PAH, are inactive as tumor initiators and tumor promoters, but accelerate the carcinogenicity and tumor initiating activity of active polycyclic hydrocarbons when given jointly. We identified trans-4,4'-dichlorostilbene, N-alkyl indoles, and N-alkyl carbazoles as tumor accelerators.

One specific characteristic of tobacco smoke is its tumor promoting activity. Until now, only limited information has been available as to the chemical nature of the tumor promoters. Although volatile phenols and long chain fatty acids are known tumor promoters when applied in high concentrations, the majority of the promoters in the "tar" remain unknown and need to be identified. These unknowns reside in the acidic portion of the particulate matter.

Despite the fact that trace amounts of known bladder carcinogens have been found in tobacco smoke, it appears more likely that tobacco smoke induces changes in the metabolism of tryptophan, which may be related to the increased risk of cigarette smoke to develop bladder cancer.

REFERENCES

1. Wynder, E.L. and Hoffmann, D.: Tobacco and Tobacco Smoke.
    Studies in Experimental Carcinogenesis. Academic Press
    Inc., New York, 1967, 730pp

2. U.S. Public Health Service: The Health Consequence of Smoking.
    A Report of the Surgeon General; 1971, U.S. Department
    of Health, Education, and Welfare, Washington, D.C.488pp

3. Hoffmann, D. and Wynder, E.L.: A study of tobacco carcinogene-
    sis. XI. Tumor initiators, tumor accelerators, and tumor
    promoting activity of condensate fractions. Cancer, 27:
    848-864, 1971

4. Otto, H. and Elmenhorst, H.: Experimentelle Untersuchungen
    zur Tumorinduktion mit der Gasphase des Zigarettenrauches.
    Z.Krebsforsch., 70:45-47, 1967

5. Leuchtenberger, C. and Leuchtenberger, R.: Effects of chronic
    inhalation of whole fresh cigarette smoke and of its gas
    phase on pulmonary tumorigenesis in Smell's mice. In
    U.S. At. Energ. Symp. Ser., 21:329-346, 1970

6. Harke, H.P., Drews, C.J., and Schüler, D.: Über das Vorkommen
    von Norborenderivaten in Tabakrauchkondensaten. Tetra-
    hedron Letters No. 43:3789-3790, 1970

7. Neurath, G.: Zur Frage des Vorkommens von N-Nitroso-Verbindun-
    gen im Tabakrauch. Experientia, 23:400-404, 1967

8. Holland, R.H. and Acevedo, A.R.: The carcinogenicity of in-
    haled arsine and triphenyl arsine in rabbits. Proc.
    Amer.Assoc. Cancer Res., 5:28, 1964

9. Sunderman, F.W. and Donnelly, A.J.: Studies of nickel carcino-
    genesis. Metastasizing pulmonary tumors in rats induced
    by the inhalation of nickel carbonyl. Am.J. Path., 46:
    1027-1041, 1965

10. Deichmann, W.B., MacDonald, W.E., Lampe, W.E., Dressler, I.
    and Anderson, W.A.D.: Nitro olefins as potential carcino-
    gens in air pollution. Ind.Med.Surg., 34:800-807, 1965

11. Magee, P.N. and Barnes, J.M.: Carcinogenic nitroso compounds.
    Advan. Cancer Res., 10:163-246, 1966

12. Druckrey, H., Preussmann, R., Ivankowic, S., and Schmähl, D.:
    Organotrope, carcinogene Wirkungen bei 65 verschiedenen

N-Nitroso-Verbindungen an BD-Ratten. Z.Krebsforsch.,69:
103-201,1967

13. Newsome, J.R. and Keith, C.H.: Variation of the gas phase
    composition within a burning cigarette. Tobacco Sci.,
    9:65-69, 1965

14. Holland, R.H. and Acevedo, A.R.: Current status of arsenic
    in American cigarettes. Cancer, 19:1248-1250,1966

15. Stedman, R.L.: The chemical composition of tobacco and to-
    bacco smoke. Chem. Rev., 68:153-207, 1968

16. Day, J.M., Bateman, R.C., and Coghill, E.C.: Determination of
    trace amounts of nickel in tobacco by neutron activation
    analysis. Abstr. 145, Natl. Meet.Am.Chem.Soc., New York,
    N.Y. 23A, 1963

17. Pailer, M. and Kuhn, H.: Kurzer Bericht über das Vorkommen
    von Nickel im Zigarettenrauch. Fachl. Mitt.Oesterr.
    Tabakregie , 4:61-63, 1963

18. Fresh, J.W., Sun, S.C. and Rampsch, J.W.: Nasopharyngeal car-
    cinoma and environmental carcinogens. In Unio Intern.
    Contra Cancrum Monogr.,1:124-129, 1967

19. Sunderman, F.W. and Sunderman, Jr. F.W.: Nickel poisoning.
    XI. Implication of nickel as a pulmonary carcinogen in
    tobacco smoke. Am.J.Clin.Pathol., 35:203-209, 1961

20. Bartle, D.D., Bergstedt, L., Novotny, M., and Widmark, G.:
    Tobacco chemistry. II. Analysis of the gas phase of to-
    bacco smoke by gas chromatography-mass spectrometry. J.
    Chromatog., 45:256-263, 1969

21. Hoffmann, D. and Rathkamp, G.: Chemical studies on tobacco
    smoke. III. Primary and secondary nitroalkanes in ciga-
    rette smoke. Beitr. Tabakforsch., 4:124-134, 1968

22. Hoffmann, D. and Rathkamp, G.: Chemical studies on tobacco
    smoke. XII. Quantitative determination of nitrobenzenes
    in cigarette smoke. Anal. Chem., 42:1643-1647, 1970

23. Kallianos, A.G., Means, R.E., and Mold, J.D.: Effect of ni-
    trates in tobacco on the catechol yield in cigarette
    smoke. Abstr. 20th Tobacco Chemists'Res. Conf. 1965,
    21-22, Winston-Salem, N.C. ; Tobacco Science, 12:125-
    129, 1968

24. Neurath, G., Pirman, B., and Wichern, H.: Zur Frage der N-
     Nitrosoverbindungen im Tabakrauch. Beitr. Tabakforsch.,
     2:311-319, 1964

25. Serfontein, W.J. and Smit, J.H.: Evidence for the occurrence
     of N-nitrosamines in tobacco. Nature, 214:169-170, 1967

26. Johnson, D.E., Millar, J.D., and Rhoades, J.W.: Nitrosamines
     in tobacco smoke. Natl. Cancer Inst. Monogr. 28:181-189,
     1969

27. Pailer, M. and Klus, H.: Die Bestimmung von N-Nitrosaminen
     im Zigarettenrauchkondensat. Fachl. Mitt. Oesterr.Tabak-
     regier, 12:203-211, 1971

28. Neurath, G: Stickstoffverbindungen des Tabakrauches. Arznei-
     mittel-Forsch., 19:1093-1106, 1969

29. Norman, V. and Keith, C.H.: Nitrogen oxides in tobacco smoke.
     Nature, 205:915-916, 1965

30. Hoffmann, D. and Vais, J: Analysis of volatile N-nitrosamines
     in unaged mainstream smoke of cigarettes. Pres. at the
     25th Tobacco Chemists' Research Conf., 1971

31. Cooper, J.T., Forbes, W.F., and Robinson, J.C.: Free radicals
     as possible contributors to tobacco smoke carcinogenesis.
     In Natl. Cancer Inst. Monogr.,28:191-197, 1971

32. Boening, H.V.: Investigation of the Pyrolytic Products of Irri-
     diated Cigarettes. Spindletop Research, Lexington, Ky.,
     1965, 33pp

33. Boening, H.V.: Free Radicals and Health. Spindletop Research,
     Lexington, Ky., 1966, 133pp

34. Bluhm, A.L., Weinstein, J., and Sousa, J.A.: Free radicals in
     tobacco smoke. Nature, 229:500, 1971

35. Peacock, P.R., and Spence, J.B.: Incidence of lung tumors in
     LX mice exposed to (1) free radicals, (2) SO_2. Brit. J.
     Cancer, 21:606-618, 1967

36. Wynder, E.L., Graham, E.A., and Croninger, A.B.: Experimental
     production of carcinoma with cigarette tar. Cancer Res.
     13:855-864, 1953

37. Dontenwill, W., Elmenhorst, H., Harke, H.P. Reckzeh, G., We-
     ber, K.H., Misfeld, J., and Timm, J.: Experimentelle Un-

tersuchungen über die tumorerzeugende Wirkung von Zi-
garettenrauch-Kondensaten an der Mäusehaut. Z.Krebs-
forsch., 73:265-314, 1970

38. Whitehead, J.K. and Rothwell, K.: The mouse skin carcinogenicity of cigarette smoke condensate: Fractionated by solvent partition methods. Brit.J. Cancer, 23:840-857, 1969

39. Bock, F.G., Swain, A.P., and Stedman, R.L.: Bioassay of major
fractions of cigarette smoke condensate by an accelerated technic. Cancer Res., 29:584-587,1969

40. Wynder, E.L. and Hoffmann, D: A study of tobacco carcinogenesis. X. Tumor promoting activity. Cancer, 24:289-301,1969

41. Wynder, E.L. and Hoffmann, D.: The epidermis and the respiratory tract as bioassay systems in tobacco carcinogenesis.
Brit.J. Cancer, 24:574-587, 1970

42. Wynder, E.L. and Wright, G.L A study of tobacco carcinogenesis. I. The primary fractions. Cancer, 10:255-271,1957

43. Seelkopf, C., Ricken, W., and Dhom, G.: Untersuchungen über
die krebserzeugenden Eigenschaften des Zigarettenteeres.
Z.Krebsforsch., 65:241-249, 1963

44. Borinsynk, Y.P.: Blastomogenic activity of smoking products.
Vop.Eksp.Onkol. No. 3:47-55, 1968

45. Hoffmann, D. and Wynder, E.L.: The tumor initiators in tobacco
smoke. Proc.Amer.Assoc. Cancer Res., 7:32, 1966

46. Hoffmann, D. and Wynder, E.L.: Selective reduction of the tumorigenicity of tobacco smoke. Experimental approaches.
In Natl. Cancer Inst. Monogr., 28:151-172, 1968

47. Hoffmann, D. and Wynder, E.L.: Selective reduction of the
tumorigenicity of tobacco smoke. Experimental approaches.
II. Presented at the 2nd World Conference on Smoking
and Health, London, September 1971

48. Hoffmann, D., Rathkamp, G., and Wynder, E.L.: Chemical studies
on tobacco smoke. V. Quantitative determination of chlorinated hydrocarbon insecticides in cigarette tobacco and
its smoke. Beitr. Tabakforsch., 4:201-214, 1968

49. Hoffmann, D., Rathkamp, G., and Nesnow, S.: Chemical studies
on tobacco smoke. VIII. Quantitative determination of
9-methylcarbazoles in cigarette smoke. Anal. Chem., 41:
1256-1259, 1969

50. Hoffmann, D., Rathkamp, G., Nesnow, S., and Wynder, E.L.:
     Chemical studies on tobacco smoke. XVI. Methylfluoran-
     thenes identification in cigarette smoke and tumor ini-
     tiating activity. In prep.

51. Chan, P.: Unpublished data, 1971

52. Gellhorn, A.: Cocarcinogenic activity of cigarette tobacco
     tar. Cancer Res., 18:510-517, 1958

53. Roe, F.J.C., Salaman, M.H., and Cohen, J.: Incomplete carcino-
     gens in cigarette smoke condensate; tumor promotion by
     phenolic fraction. Brit. J. Cancer, 13:623-633, 1959

54. Van Duuren, B.L., Sivak, A., Katz, C., and Melchion, S.: Ci-
     garette-smoke carcinogenesis. Importance of tumor pro-
     moters. J. Natl.Cancer Inst., 47:235-240, 1971

55. Bock, F.G.: Methods for bioassays of tobacco smoke. Pres.
     Symposium "Composition of Tobacco Smoke", Div. Agr.Food
     Chem., 162nd Natl. Meet.Am.Chem.Soc., Washington, D.C.
     Sep. 12-17, 1971

56. Hoffmann, D. and Woziwodzki, H.: Chemical studies on tobacco
     smoke. IV. Quantitative determination of free non-vola-
     tile fatty acids in tobacco and tobacco smoke. Beitr.
     Tabakforsch., 4:167-174, 1968

57. Holsti, P.: Tumor promoting effect of some long chain fatty
     acids in experimental skin carcinogenesis in the mouse.
     Acta Pathol Microbiol. Scand., 46:51-58, 1959

58. Muñoz, N., Correa, P. and Bock, F.G.: Comparative carcinogenic
     effect of two types of tobacco. Cancer, 21:376-389, 1968

59. Pailer, M. Hübsch, W.J., and Kuhn, H.: Untersuchungen der ali-
     phatischen und armatischen primären und sekundären Amine
     des Zigarettenrauches mit Hilfe der Gas Chromatographie
     und Massenspektrometrie. Fachl. Mitt. Oesterr. Tabakre-
     gier No. 7:109-118, 1967

60. Miller, R.L. and Stedman, R.L: Essential absence of β-naph-
     thylamine in cigarette smoke condensate. Tobacco Sci.,
     11:111, 1967

61. Masuda, Y. and Hoffmann, D.: Chemical studies on tobacco smoke.
     VII. Quantitative determination of 1-naphthylamine and
     2-naphthylamine in cigarette smoke. Anal. Chem., 41:650-
     652, 1969

62. Boyland, E.: The possible carcinogenic action of alkaloids
          of tobacco and betel nut. Planta Medica Suppl., 11:13-
          23, 1968

63. Boyland, E.: The Biochemistry of Bladder Cancer. C.C. Thomas
          Springfield, Ill., 1963, 95pp

64. Boyland, E., Busby, E.R., Dukes, C.E., Grover, P.L., and
          Manson, D.: Further experiments on implantation of
          materials into the urinary bladder of mice. Brit.J.
          Cancer, 18:575-581, 1964

65. Kerr, W.K., Barkin, M., Levers, P.E., Woo, S.K.C., and
          Menczyk, Z.: The effect of cigarette smoking on bladder
          carcinogens in man. Can.Med. Ass.J., 93:1-7, 1965

66. Brown, R.R., Price, J.M., Satter, E.J., and Wear, J.B.: The
          metabolism of tryptophan in patients with bladder can-
          cer. Am. Indust.Hyg.Assoc.J., 30:27-34, 1969

67. Hecker, E.: Über die Wirkstoffe des Crotonöls. I. Biolo-
          gische Teste zur quantitativen Messung der entzündlichen
          cocarcinogenen und toxischen Wirkung. Z.Krebsforsch.,
          65:325-333, 1963

This study was supported by the American Cancer
Society Grant BC -56P and by the National Cancer
Institute Grant NIH-NCI 70-2087

MODIFICATION OF TOBACCO SMOKE

C. H. Keith

Celanese Fibers Company

Box 1414, Charlotte, N. C. 28201

Over the years, a growing knowledge about the properties of tobacco smoke, particularly that generated by cigarettes, has led to the development of greatly modified smoking products. Some twenty years ago most U.S. cigarettes were short, unfiltered, and contained relatively unflavored robust blends of natural tobaccos. Nowadays a wide variety of filtered, ventilated, longer and some- times slimmer cigarettes are available to the smoking public. Their preference for such modified cigarettes is amply demon- strated by the growth of the filter cigarette portion from practically nothing to 80% or more of the U.S. cigarette market. A further demonstration of such modification has been the steady decrease in tar yield from cigarettes over the years. For example, the average values for popular brands decreased from 30-35 milli- grams to 18-22 milligrams per cigarette in the past 17 years.

It is the purpose of this paper to describe the various methods of modifying tobacco smoke, and to illustrate what effects these methods have on the ultimate product, the chemically complex smoke stream coming out of a cigarette. This review will hopefully provide guidelines to further modifications in tobacco products.

To understand the various processes involved in the smoking of a cigarette, it is useful to consider the first figure. In this simple illustration of a smoldering cigarette, we have depicted four zones of activity and a number of arrows indicating the flow of air into and materials out of the cigarette.

In zone A, which can be called the combustion zone, the charred residue of the tobacco filler burns in air entering through the outer layers of ash on the cone of the cigarette.

149

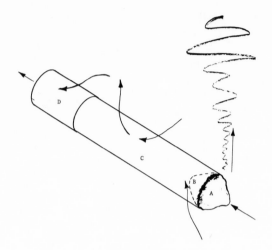

FIGURE 1

FIGURE 2

Gaseous Composition of Cigarette Smoke,
Second Puff (7.5 mm burned)
Ref. 6

Distance from cone (mm)	Volume percent of						
	N_2	O_2	CH_4	CO	H_2	CO_2	Other
0	53.5	1.4	1.3	11.8	8.2	17.7	6.1
5	53.4	2.7	1.5	10.6	6.4	22.2	3.2
20	61.7	8.6	0.8	6.4	4.5	16.6	1.4
37.5	64.9	11.1	0.5	4.9	2.3	13.1	3.2
77.5	69.4	15.5	0.3	2.4	1.2	7.6	3.6

This true combustion is the heat source for the whole smoking process, and is thought to result in relatively simple products such as CO, CO_2, H_2O, etc. This high temperature combustion region also uses up the oxygen in the entering gas stream leaving a reducing atmosphere in region B, and has a relatively high pressure drop because of thermal effects (1). This utilization of available oxygen and high pressure drop are extremely important in determining the properties of the final smoke stream. As Egerton (2) noted, the plugging effect of the hot coal causes the burning during a puff to occur mainly along the outer perimeter of the cigarette, while the slower burning during the resting interval between puffs occurs primarily in the high temperature center of the zone to re-establish the natural cone profile. Thus the sidestream smoke formed during the interval is somewhat different from that formed during the puff, being richer in high temperature species such as polycyclic hydrocarbons (3).

Two types of modifications are possible in this region of the cigarette, one being the use of additives to change the burning temperature of the cone and another to change the properties of the ash to allow more or less air to penetrate to the cone. Since the cone temperature is determined by the balance between the rate of combustion and rate of radiant heat loss to the surroundings, very large changes do not appear to be feasible. Moderate increases in the combustion temperature can be achieved by increasing the reflectivity or the insulating properties of the ash. This change generally results in a faster burning cigarette which achieves a reduction in tar levels by allowing fewer puffs under a standardized smoking routine. Materials such as sulfur, magnesium carbonate and vanadium pentoxide either in admixture with tobacco or incorporated in tobacco sheet have been reported as affecting cone temperature (4). Reducing the cone temperature has long been suggested as a means of reducing polycyclic hydrocarbons in tobacco smoke (5, 6), but, unfortunately, the heat balance does not allow a very great reduction before the cigarette goes out. Similar effects are achieved by changing the density and tenacity of the ash, a dense ash providing more insulation but also hindering the air flow into the combustible material.

A second zone of great importance is the pyrolysis zone labeled B in the figure. This area consists of tobacco which is subjected to high temperatures because of its proximity to the burning coal. As was predicted by Hobbs (1) and as shown in the data (Figure 2) of Newsome and Keith (6), the gas composition in this region is rich in hydrogen and almost devoid of oxygen. Under these conditions, pyrolytic reactions occur which form the great variety of smoke components not found in the initial tobacco. This region is also the initial point of

formation of the smoke as we know it, since the higher boiling
pyrolysis products begin to condense on nuclei produced by the
explosive heating of the combusting material. This condensation
occurs initially here because of the rapid cooling of the smoke
as the stream moves away from the hot combustion zone. As shown
by temperature profiles (7), the thermal gradient is very steep
in this region, and this results in particle concentrations of
3-5 billion particles per cc (8) which are quite uniform in size
and composition.

Altering the thermal gradient would be expected to considerably alter the smoke composition as the thermal reactions would
be more or less rapidly frozen out. A number of possible modifications have been proposed such as the addition of materials
which endothermically decompose at the applicable temperatures
and materials which fuse and hinder the passage of pyrolysis
products out of this region. Baxter and Hobbs (9) found that
carbon dioxide generated by decarboxylation of tobacco accounts
for about one half of that material in the smoke stream, and that
this can cause a quenching of the puff. Stedman, Schmeltz, and
Burdick and co-workers have investigated a number of additives to
tobacco in model pyrolysis experiments and in actual smoking
experiments. Some of these have significant effects on the composition of the smoke stream as summarized in another paper in this
Symposium and by Schmeltz (10) at the 5th CORESTA Congress.
Another pyrolytic modification which has been shown to be
effective by Schur and Rickards (11) and Rickards and Owens (12)
is to modify the paper wrapper by changing its burning characteristics and porosity. By using a more porous and/or faster
burning paper, the amount of air entering around and just behind
the cone and pyrolysis zones can be considerably changed. For
example, Newsome and Keith (5) showed from calculations based on
their gas phase measurements that the gas phase of the second puff
from a cigarette wrapped in a medium porosity paper consisted of
30% combustion gases coming from the cone and pyrolysis zones,
15% diluting air entering around the cone, and 56% diluting air
entering through the paper wrapper. Similar figures for a low
porosity paper were 36% combustion gases, 24% dilution around the
cone, and 40% dilution through the paper. A further modification
can be achieved by placing discrete rows of perforations along
the paper. These tend to provide small jets of air which penetrate
and dilute the gases in the pyrolysis zone to a greater extent
than those from a cigarette wrapped in an equivalent porosity
uniformly porous paper. One result was a large reduction in both
particulate and vapor phase components in the smoke stream, as is
found in all porous wrappings. Another and more important result
was a selectively greater reduction of gas phase constituents
such as CO, HCN, acrolein, isoprene, etc. than of particulate
matter. Higher molecular weight materials which are able to

distill into the smoke stream were not selectively reduced in these cigarettes, i.e., their yields were lowered only about as much as that of particulate matter.

As indicated in the foregoing discussion, a number of materials directly distill into the smoke stream from the mass of unburned tobacco in region C of the cigarette. Among these are well-known constituents such as nicotine and menthol, which have been shown to distill largely unchanged into the smoke stream, although the former evidently goes from a salt to a free base back to a salt in the final smoke stream. Compounds with rather low volatility are found to distill essentially unchanged in the burning cigarette environment. An example of such a material is dotriacontane, which was investigated in radiotracer experiments (13). It was found that 95% of this relatively high boiling material was transferred unchanged to the smoke stream. The remaining 5% was found to be mostly low molecular weight pyrolysis products, and practically none of the original radioactivity was found in CO and CO_2. These findings suggest a clean-cut distillation process with a minor fraction of the material being subjected to a vigorous pyrolytic degradation. In passing, it should be again recognized and emphasized, as was done by Dawson (14), that such radioactive tracer experiments are the best existing technique for truly understanding the chemistry of the smoking process, and many experiments are needed in this area.

Modification of the smoke stream in this distillation zone is usually achieved by changing the tobacco composition and/or the dilution process. Extraction of tobacco, use of different tobaccos, changing the constituents of reconstituted tobaccos, and the use of non-tobacco components in such materials or by themselves will certainly alter the amount of distillable components in the smoke stream. Similarly physical modifications such as puffing the tobacco, as described by Johnson (15), or making a low density reconstituted sheet (16) alter the concentrations of distillable components in the filler, and, consequently, in the smoke stream. These expanded filler materials also change the combustion and pyrolysis properties of the cigarette, usually resulting in a rather fast burning smoking article.

Increasing the porosity of the paper wrapper also changes the concentrations of the distillable smoke components by diluting the smoke stream with external air (11). A companion effect studied by Owen and Reynolds (17) is the diffusion of light gases out of the cigarette in this region. Considerable losses of hydrogen through this outward diffusion were noted by us (5), and were quantified by them.

The final point for modification of the smoke stream exists
in the filter. This is the most commonly used means of altering
the composition and quantity of smoke coming from a cigarette.
In its most common form, a bundle or tow of ten to fifteen
thousand crimped, continuous cellulose acetate filaments, it has
received wide public acceptance. Approximately 80% of all cigarettes in the U.S. are equipped with this type of filter, and its
usage throughout the world is growing towards that figure. The
primary action of a filter is to remove matter from the smoke
stream by mechanical capture of smoke particles. Depending on
the structure and composition of the filter, it can remove from
20 to 75% of this particulate matter, and most commonly the
removal efficiency lies between 40 and 55%. A companion property
to removal efficiency for any filter system is its draw resistance
or pressure drop. This is a measure of the suction which must be
applied to withdraw smoke at a fixed flow rate, which is generally
chosen as the average puff velocity, 17.5 milliliters per second.
Obviously, the most effective filter is that which has a high
removal efficiency at a low pressure drop, usually around 60 mm
of water.

The mechanisms of particulate filtration and pressure drop
generation have been investigated both theoretically and experimentally. Excellent agreement between theory and experiment has
been achieved in both instances in work by Keith, Reynolds, and
Dalton (18, 19, 20, 21). Pressure drop is generated by the drag
of the stationary fibers on the moving smoke stream. Factors such
as the number of fibers, fiber orientation, fiber surface area,
and degree of fiber touching or opening greatly influence pressure
drop.

Particulate filtration has been found to largely result from
Brownian diffusion of particles to the filter strands and direct
interception of particles by appropriately placed fibers. An
estimate of the contribution of these two mechanisms and the less
important inertial impaction mechanism has been given, this being
66% diffusion, 33% interception, and 1% impaction (21). The most
important variables in controlling the mechanical removal of
particles, as in the case of pressure drop, are the number and
size of fibers and the interfiber separation. Fiber orientation
and cross section are of much lesser importance in this process,
and the nature of the filtering material is unimportant in this
purely mechanical process as the capture of particles appears to
be irreversible (22). As will be discussed later, the chemical
nature of the filtering material plays an important role in
determining the removal of volatile and semi-volatile smoke components. It thus largely controls the ability of the filter to
operate selectively for some of these components.

FIGURE 3

SMOKE REMOVAL EFFICIENCY AS A FUNCTION
OF PRESSURE DROP AND FILTER LENGTH

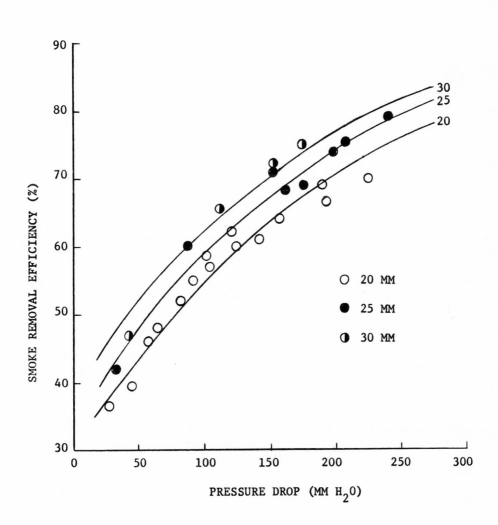

Experimentally, acetate filters have been found to operate over a wide range of smoke removal efficiencies or SRE, which is defined as the weighable material captured by the filter divided by the total weighable material entering the filter under a standard smoking routine such as that used by the Federal Trade Commission. This range of performance is illustrated in Figure 3, where the SRE values for a variety of acetate filters of ordinary construction are plotted as a function of pressure drop and filter length.

Practically, the variables of importance in controlling removal efficiency for ordinary acetate filters are found to be filter length, pressure drop, circumference, and fiber denier. Other variables such as fiber weight are of importance but their contribution is included in the pressure drop effect. Empirically determined equations relating these variables are given in Figure 4 for SRE and for similarly defined nicotine and tar removal efficiencies, tar being defined as weighable material collected on a Cambridge filter pad with water and nicotine subtracted therefrom. These empirical equations are of considerable practical value in filter design, and they show that longer, higher pressure drop, larger circumference, finer fiber filters are the more efficient variants.

As has been illustrated, pressure drop and filtration efficiency are closely related. As one goes up so does the other. However, this is not always the case, as the filter structure can be changed. One approach has been to create a non-uniform distribution of fibers by introducing a number of channels or corrugations running the length of the filter. These channels provide a pressure drop relief, and, if the pressure drop along the channel walls is balanced, the efficiency is not significantly diminished. Such corrugated structures are commonly found in crepe paper filters which are used in some dual filter structures and have been successfully employed in acetate and other polymer filters. The general approach is to use an extremely small or irregular fiber to achieve high filtration efficiency and to reduce the pressure drop to acceptable levels by providing the channel structure.

Another approach that is commercially utilized is to increase the face area of the filter by having the smoke flow through the relatively dense walls of a tubular shaped filter structure (23). A high filtration is achieved by the close packing of the fine filter fibers, but pressure drop is kept at a reasonable level by the increased area of the filter presented to the stream.

A third method of controlling pressure drop by altering the filter structure is to provide ventilation through the walls of

FIGURE 4

$$\log_e \left(1 - \frac{E}{100}\right) = A \cdot L + B \cdot \Delta P \cdot C^4 + D \cdot L/\delta$$

Where   E = Removal Efficiency (Smoke, Nicotine, or Tar)

        L = Filter Length in MM

      ΔP = Filter Pressure Drop in MM H_2O at 17.5 ml/sec. Flow

      C = Filter Circumference in MM

      δ = Fiber Denier Per Filament in gm.

      A, B, and D = Constants from table below.

	A	B	D
Smoke:	-1.542×10^{-2}	-9.602×10^{-9}	-2.102×10^{-2}
Nicotine:	-3.822×10^{-3}	-1.048×10^{-8}	-1.824×10^{-2}
Tar:	-9.957×10^{-3}	-8.517×10^{-9}	-2.587×10^{-2}

the cylindrical filter. This can either be a controlled addition
of air into the mass of filtering material or it can be arranged
so that the air is introduced at a mouthpiece section behind the
filter (24, 25). Both types provide a reduction in pressure drop,
and, because less smoke has to be withdrawn from the cigarette to
form a 35 ml puff, a considerable reduction in tar delivery is
also achieved. A parallel effect is that the smoke is drawn
through the filter at a slower rate, which is favorable for more
efficient filtration. One effect of air dilution within the
filter, but not that occurring after the filter, is a change in
the vapor balance in the smoke stream. The diluting air can sweep
previously deposited volatile smoke components off the filter
material, thereby enhancing their concentration in the effluent
stream.

Another type of filter system which can be effective as a
mechanical filter consists of a loose bed or foam structure of a
granular or porous solid (26). Usually this filter bed is contained in a cavity or chamber with one or more ordinary fiber
filter segments adjacent to it for mechanical and aesthetic reasons.
In such filters, the important factors controlling the filtration
efficiency are, as in the case of fibrous structures, the surface
area available to the smoke stream and the spacing between the
granules. Pores in the granular or foam material assist in the
filtration process provided that they are sufficiently large to
permit the smoke particles to freely enter.

One type of granular filtering material which has found quite
wide public acceptance is activated charcoal. This material is
primarily useful in removing gaseous or vapor phase components
from the smoke stream since the pore size is not sufficiently
large for effective particle removal. Since it is primarily a
vapor phase filtering agent, it may be considered to be a
selective filter for these materials.

Selective filtration, which is defined as the ability of a
filter to remove more or less of a given component than the smoke
stream as a whole, is evident in a variety of filters. The
mechanism appears from all available studies to be one of vapor
phase adsorption and desorption of components with reasonable
volatilities. Because of this, components with very low and very
high vapor pressures are not susceptible to selective removal.
In the case of very low vapor pressures, the material is entirely
contained in the smoke particles and is removed by the non-
selective, irreversible mechanical filtration process. At the
other end of the volatility scale, gaseous materials with very
high vapor pressures do not appreciably condense on the filtering
material, so that there is relatively little opportunity for the
filter to selectively extract them from the smoke stream.

Materials with intermediate volatility can and do frequently contact the filter material during their brief transit through the filter. If there is a degree of chemical and physical affinity between these materials and the filter substrate, then the sorption and desorption of the material is altered and there is a selective retention of the material and a reduction of its concentration in the smoke stream. Similarly, a material with a low affinity for the filter material is preferentially enriched in the smoke stream and can be said to be undergoing reverse selectivity. Since volatile and semi-volatile materials are the major contributors to the taste and aroma of tobacco smoke, it is obvious that the selective removal properties of a filter will cause considerable alteration of the taste of the smoke stream. The public acceptance or rejection of cigarettes is largely dependent on this factor and many technically feasible and otherwise acceptable filters have been discarded because of a serious distortion in the taste spectrum.

A convenient means of describing selectivity is through dimensionless separation factors such as the Davis and George (27) selectivity factor. This is computed by dividing the concentration of a given component in unfiltered smoke by that in filtered smoke. Thus a selective removal would be indicated by a dimensionless number greater than 1, a non-selective removal by a number close to 1, and reverse selectivity by a fraction between 0 and 1.

Although any filtering material has a degree of selectivity for some smoke components, two are of particular interest because of their commercial usage and because they illustrate different mechanisms of selective removal. These are activated carbon and cellulose acetate.

In the case of carbon and other adsorbents, the smoke vapors are physically plated onto an extensive but relatively impermeable substrate. The surface can be treated to chemically react with particular vapors, as, for example, coating the surface with metallic oxides to improve the selectivity of charcoal for acidic gases (28). However, the adsorbed molecules do not further penetrate into the substrate, and the adsorption capacity is dictated primarily by the surface area of the charcoal. Since this surface is very large - being of the order of 1000 to 1200 sq.meters per gram - quite appreciable selective removals can be achieved. This is illustrated in Figure 5, showing selectivity factors for acrolein - a representative fairly volatile smoke component - for a variety of carbon filters. These filters are arranged in approximate order of increasing carbon content, and as would be expected, they show increasing selectivity for this irritating smoke component.

FIGURE 5

SELECTIVITY OF VARIOUS CHARCOAL FILTERS FOR ACROLEIN

FILTER	SELECTIVITY FACTOR FOR ACROLEIN	DATA SOURCE
PAPER + CARBON - 7.5 MM[1]	1.4	IRBY [29]
PAPER + CARBON - 7.5 MM[1]	1.0	WILLIAMSON[30]
PAPER + CARBON - 15 MM	1.6	WILLIAMSON[30]
ACETATE + CARBON - 7.5 MM[1]	1.3	WILLIAMSON[30]
ACETATE + CARBON - 15 MM	2.2	WILLIAMSON[30]
LOOSE CARBON BED - 5 MM[2],[3]	2.3	NEWSOME[31]
BONDED CARBON PLUG - 7.5 MM[1]	6.5	WILLIAMSON[30]
BONDED CARBON PLUG - 10 MM[3][4]	3.5	LAURENE[32]
BONDED CARBON PLUG - 15 MM	10.6	WILLIAMSON[30]

(1)   BACKED BY 7.5 MM ACETATE SECTION

(2)   RESTRAINED BY TWO 7.5 MM ACETATE PLUGS

(3)   ESTIMATED FROM INDEPENDENT MEASUREMENTS OF PARTICULATE
       MATTER FROM EQUIVALENT CIGARETTES

(4)   BACKED BY 10 MM ACETATE SECTION

FIGURE 6

SELECTIVITY DATA FOR PHENOL

DESCRIPTION OF FILTER	SELECTIVITY FACTOR FOR PHENOL	DATA SOURCE
POLYETHYLENE	0.8	SPEARS [33]
POLYPROPYLENE	1.0	SPEARS [33]
CELLULOSE PAPER	0.9	SPEARS [33]
	1.3	WALTZ [34]
	1.0	WILLIAMSON [30]
	1.1	LIPP [35]
VISCOSE	1.1	WILLIAMSON [30]
PAPER & CARBON	1.1	WILLIAMSON [30]
BONDED CARBON	1.0	WILLIAMSON [30]
ACETATE + CARBON	1.3	WILLIAMSON [30]
CELLULOSE ACETATE – UNPLASTICIZED	1.9	DAVIS [27]
	1.9	WILLIAMSON [30]
PLASTICIZED	1.6–2.1	HOFFMAN [36]
	1.8–2.6	SPEARS [33]
	1.8–2.1	LIPP [35]
	2.6	WALTZ [34]
	2.8	DAVIS [27]
HIGHLY PLASTICIZED	4.1	DAVIS [27]
CELLULOSE TRIACETATE – PLASTICIZED	2.1	SPEARS [33]
	2.4	DAVIS [27]
	2.4	HOFFMAN [36]
PLASTICIZED, GRAINY SURFACE	3.6	HOFFMAN [36]

Another type of selective removal is found to occur in cellulose acetate filters. In this type of filter, the surface area
is less extensive, being on the order of .25 square meters per
gram; yet quite selective removal of certain smoke components can
be achieved. The reason is that the surface is acting as an
absorbent in that the condensed material is dissolved in the body
of the fiber, thereby allowing further deposition on the surface.
This is demonstrated in Figure 6 by selectivity data for
phenol on a variety of filtering materials. Of the considerable
variety of materials tested, only cellulose acetate and cellulose
triacetate are found to selectively remove phenol. It is also
apparent in these data that the addition of plasticizing or
softening agents improves this selectivity for this irritating
smoke component. It is thought that a surface coating of these
plasticizing materials serves as the primary absorbent for particular smoke components, and they also promote the diffusion of captured molecules into the body of the fiber.

This diffusion effect is further demonstrated by the data in
Figure 7   (37). In these experiments, a common plasticizing
material was applied to several permeable and impermeable fibrous
substrates and the removal efficiencies measured for several
chemically dissimilar volatile smoke components. In these long,
heavily loaded filters, the nature of the fibrous substrate has a

FIGURE 7

Vapor Removal Efficiencies of Various
Fibers Treated with Triethyl Citrate
(44-50% added, 50 mm tips)
Ref. 37

Fiber	Isoprene	Acetaldehyde	Methanol	Acetone	Toluene
Cellulose Acetate	20	38	75	60	79
Polypropylene	30	31	68	54	86
Glass Wool	0	3	2	38	72
Rayon	18	20	36	36	76

definite effect on the removal efficiency. Inert, impermeable
fibers such as glass wool and rayon have relatively little effect
on the lighter, more volatile components. When a more permeable
substrate, such as polypropylene or cellulose acetate, is provided,
the removal efficiencies increase markedly. It also is evident
that the chemical nature of the smoke component and the filter
material is important, the cellulose acetate substrate being more
effective for polar materials and the hydrocarbon polymer being
more effective for the chemically similar non-polar components.

This and other data indicate that selective filter systems
can be devised, and these can considerably alter the composition
of the vapor phase of the smoke stream.

In summary, it has been shown that tobacco smoke, principally
that coming from cigarettes, can be purposely altered by a variety
of techniques. The smoke stream issuing from a cigarette can be
changed by:

1. Altering the primary combustion process by means of additives
    and cigarette structure.

2. Changing the dimensions, air dilution, and composition of the
    charred mass undergoing a pyrolytic degradation.

3. Altering the dilution and distillation of components from the
    unburned filler.

4. Changing the composition or physical structure of the com-
    bustible filler.

5. Providing more or less mechanical filtration.

6. Providing controlled dilution before, in, or after the filter.

7. Adding selective components to the filter system.

Most, if not all, present day cigarettes utilize combinations of
these systems to provide smoking articles substantially different
from those of years past.

References

(1) Hobbs, M. E.; Tob. Sci. 1, 7477, 1957.

(2) Egerton, A., K. Gugan, and F. J. Weinberg; Combustion Flame
     7 (1), 63-78, 1936.

(3) Wynder, E. L. and D. Hoffmann, Eds., Tobacco and Tobacco
     Smoke, 1st. Ed., Chapter 8, p. 344, Academic Press, New York,
     N. Y. 1967.

(4) Miller, F., W. J. Freeman, and R. L. Stedman, Beitr.
     Tabakforsch., 4, 269-272, 1968.

(5) Grimmer, G., A. Glasser, and G. Wilhelm, Beitr. Tabakforsch.,
     3, 415-419, 1966.

(6) Newsome, J. R. and C. H. Keith, Tob. Sci. 9, 65-69, 1965.

(7) Toyey, G. P. and R. C. Mumpower, Tob. Sci., 1, 33-37, 1957.

(8) Keith, C. H. and J. C. Derrick, J. Colloid Sci., 15, 340-356,

(9) Baxter, J. E. and M. E. Hobbs, Tob. Sci., 11, 65-71, 1967.

(10) Schmeltz, I. Paper presented at the 5th International Tobacco
     Scientific Congress, Hamburg, W. Germany, September 14-18,
     1970.

(11) Schur, M. O. and j. C. Rickards, Tob. Sci., 69-77, 1960.

(12) J. C. Rickards and W. F. Owens, Jr. Paper presented at the
     20th Tobacco Chemists Research Conference, Winston-Salem,
     N. C., November 2, 1966.

(13) Carpenter, R. D., F. L. Gager, and R. W. Jenkins, Paper
     presented at the 24th Tobacco Chemists Research Conference,
     Montreal, P.Q., October 28-30, 1970.

(14) Dawson, R. F., Paper presented at the 5th International
     Tobacco Scientific Congress, Hamburg, W. Germany, September
     14-18, 1970.

(15) Johnson, W. H., Paper presented at the 23rd Tobacco Chemists
     Research Conference, Philadelphia, Pa., October 22-24, 1969.

(16) Halter, H. M., Paper presented at the 5th International
     Tobacco Scientific Congress, Hamburg, W. Germany, September
     14-18, 1970.

(17) Owen, W. C. and M. L. Reynolds, Tob. Sci., 11, 14-20, 1967.

(18) Keith, C. H., J. O. Dalton, and M. L. Reynolds, Paper presented at the 21st Tobacco Chemists Research Conference, Durham, N. C., October 19-20, 1967.

(19) Keith, C. H., Paper presented at the 22nd Tobacco Chemists Research Conference, Richmond, Virginia, October 17-19, 1969.

(20) Reynolds, M. L., Paper presented at the 24th Tobacco Chemists Research Conference, Montreal, P.Q., October 28-30, 1970.

(21) Keith, C. H., Paper presented at the 24th Tobacco Chemists Research Conference, Montreal, P.Q., October 28-30, 1970.

(22) Keith, C. H. and P. W. Mayer, Paper presented at the 23rd Tobacco Chemists Research Conference, Richmond, Virginia, October 22-24, 1969.

(23) Berger, R. M. and E. W. Brooks, U.S. Patent No. 3,533,416, October 13, 1970.

(24) Allseits, F., and J. Doblin, U.S. Patent No. 3,396,733, August 13, 1968.

(25) Osmalov, J. O., A. R. Pacquine, R. B. Seligman, and A. C. Britton, U.S. Patent No. 3,490,461.

(26) Kay, G. I. and N. M. Bikales, Paper presented at the 22nd Tobacco Chemists Research Conference, Richmond, Virginia, October 17-19, 1968.

(27) Davis, H. and T. W. George, Beitr. Tabakforsch. 3, 203-214, 1965.

(28) Keith, C. H., V. Norman, and W. W. Bates, Jr., U.S. Patent No. 3,251,365, May 17, 1966.

(29) Irby, R. M., Jr., and E. S. Harlow, Tob. Sci., 3, 52-56, 1959.

(30) Williamson, J. T., J. F. Graham, and D. R. Allman, Beitr. Tabakforsch., 3, 233-242, 1965.

(31) Newsome, J. R., V. Norman, and C. H. Keith, Tob. Sci., 9, 102-110, 1965.

(32)   Laurene, A. H., L. A. Lyerly, and G. W. Young, Tob. Sci., 8, 150-153, 1964.

(33)   Spears, A., Tob. Sci., 7, 76-80, 1963.

(34)   Waltz, P., and M. Hausermann, Beitr. Tabakforsch., 3, 169-193, 1965.

(35)   Lipp, G., Beitr. Tabakforsch., 3, 109-127, 1965.

(36)   Hoffmann, D. and E. L. Wynder, J. Nat. Cancer Inst., 30, 67-84, 1963.

(37)   Keith, C. H. and J. R. Misenheimer, Beitr. Tabakforsch., 3, 583-589, 1966.

FILTRATION OF CIGARETTE SMOKE

J. E. Kiefer

Research Laboratories, Tennessee Eastman Company, Division

of Eastman Kodak Company, Kingsport, Tennessee 37662

INTRODUCTION

Cigarette filters began gaining acceptance in the early 1950's and their use has grown steadily since that time. In 1965, filter-tip cigarettes accounted for 46% of total world output.[1] By 1970, this figure had increased to about 60%. The increased use of cigarette filters has been accompanied by a marked change in filter performance requirements, and the quest for cigarette filters with improved performance properties has led many investigators to a study of the fundamental principles of cigarette smoke filtration.

The processes encountered in cigarette smoke filtration are somewhat similar to those encountered in the filtration of other aerosols. However, filtration of cigarette smoke presents some unique problems. For example, a cigarette filter must have a pressure drop, or resistance to draw, within closely prescribed limits, and its size and shape are severely limited by practical considerations. Also, the materials from which a filter is made must be nontoxic, tasteless, economical, readily available, and adaptable to rapid fabrication into structures that resist distortion in a smoker's mouth.

Filtration of simple aerosols is a complex phenomenon, as demonstrated by the works of Davies[2], Dorman[3], Pick[4], Langmuir[5], and many others. The complexity — both physical and chemical — of the cigarette smoke aerosol further complicates the derivation of a comprehensive theory that adequately describes the filtration process which takes place within a cigarette filter. Some of the processes that occur within a cigarette filter and some of the progress that has been made toward an understanding of the filtration process are presented in the following discussion. Emphasis is placed on developments that have been made during the past 2 to 3 years.

167

To relate general filtration theory to the special case of cigarette smoke, much literature has been published on the physical and chemical nature of both smoke aerosols and cigarette filters. For example, at the CORESTA Fifth International Tobacco Scientific Congress in Hamburg in 1970, Stöber[6] presented a very comprehensive review of the fundamental concepts of the physical chemistry of smoke aerosols.

FORMATION OF CIGARETTE SMOKE

The important processes related to the formation of cigarette smoke and the physical changes occurring within a smoke aerosol are pyrolysis, vaporization, condensation, and coagulation. According to Touey and Mumpower[7], the burning-zone temperature of a cigarette during puffing is 850 to 900°C. Pyrolysis of tobacco at this temperature produces hundreds of volatile compounds. In addition, some volatile components, such as nicotine and other alkaloids, vaporize from tobacco as a result of the heat in the burning zone. The hot gases at the burning zone contain a large number of vaporized compounds that have a wide range of boiling temperatures. This gas mixture, as it passes through the tobacco column, is cooled very rapidly. In a few hundredths of a second, the temperature drops from over 800°C to slightly above ambient temperature. These conditions are ideal for the formation of an aerosol, particularly since there are a large number of nuclei present to initiate condensation. The nuclei can be microscopic pieces of incompletely burned organic matter, carbon, ash, or other nonvolatile materials that are ejected into the gas stream from the sputtering burning zone. The existence of carbon particles and nonvolatile materials, such as sugars, in cigarette smoke indicates that these materials could be nuclei for the aerosol particles. Very large organic molecules, as well as ionized molecules, may also form the nuclei for some smoke particles. It is also possible that nucleation is not necessary for the formation of many of the particles; some may form simply by supersaturation of the cooling vapors. In any case, as Stöber[6] pointed out, condensation takes place very rapidly. This rapid temperature drop causes nearly simultaneous growth on all nuclei available, and this condition leads to the formation of a reasonably monodispersed aerosol.

As the aerosol passes through the tobacco column toward the filter, additional changes take place. Much of the condensable portion of the gas phase condenses on strands of tobacco. Buildup of moisture in the tobacco butt is evidence of this phenomenon. However, it was shown by Osdene[8] and his co-workers in 1970, that the high-boiling components in the mainstream smoke are not captured within the tobacco column. The significance of this discovery is that the smoke particles are not trapped by the tobacco strands; consequently, the filtration which occurs within the tobacco column is attributed almost entirely to the condensation mechanism.

As smoke passes through the tobacco column, it is subject to coalescence, and the very high concentration of particles in the mainstream of the smoke suggests that coagulation might be an important aspect of particle growth.

However, experimental investigations indicate that the coagulation rate of cigarette smoke is too slow for coagulation to be a major factor in particle growth. Particle growth is due mainly to condensation.

PHYSICAL PROPERTIES OF CIGARETTE SMOKE

Some of the physical properties of smoke aerosols are shown in Table I. An aerosol contains from 10^8 to 10^{10} particles/cm^3. The size of the particles varies from less than 0.1 to about 1 μ in diameter; the number-average size is 0.15 to 0.20 μ, and the mass-average size is 0.5 to 0.6 μ. These particles constitute 5 to 10% of the total weight of the smoke and contain the portion of smoke commonly referred to as tars, or total particulate matter (TPM). The linear velocity of these particles is about 35 cm/sec, which means the particles are in the filter for approximately 0.04 sec. During the first few puffs, the aerosol entering the filter is near ambient temperature, and the particles contain a high proportion of the components that have a low vapor pressure at room temperature. During the last few puffs, the smoke temperature approaches 90°C at the entrance to the filter, and many more components are at least partially in the gaseous phase of the smoke. Since the filtering action depends on the physical state of the compound being filtered, the efficiency of a filter for removing many of the components of smoke changes as the cigarette is consumed.

TABLE I
PROPERTIES OF CIGARETTE SMOKE

Property	Value
Particle concentration/cm^3	10^8 – 10^{10}
Particle size	
Range, μ	>0 but < 1.0
Number average	0.15 – 0.20
Mass average	0.5 – 0.6
Linear velocity, cm/sec	35
Residence time, sec	0.03 – 0.06
Temperature, °C	30 – 90

PROPERTIES OF CIGARETTE FILTERS

The action of the filter depends on its physical and chemical nature. Many types of filters have been disclosed in patent literature and elsewhere; however, the filters that have gained widespread acceptance are composed of fibrous materials, adsorbents, or both. Most commercial filters are made from paper or cellulose acetate fibers, whereas most adsorbent-type filters contain carbon. Combinations of fibrous materials and carbon are also used extensively.

The physical nature of a typical fibrous filter is illustrated in Figure 1. Fibers in this filter are about 20 µ in diameter, and the average distance between them is about 25 µ. A 20-mm filter of this construction has a surface area of about 27,000 mm². In comparison, the diameter of a typical smoke particle is about 0.2 µ or one-hundredth that of the fiber.

FIGURE I
TYPICAL CIGARETTE FILTER

MECHANISMS FOR FILTRATION

As the smoke aerosol passes through the filter, filtration can take place by any of three mechanisms: (1) mechanical trapping of particles, (2) condensation followed by adsorption from the gas phase of smoke, or (3) transfer between the particle and the filter. Mechanical filtration is the removal of a particle due to a collision between it and the filter surface. This type of filtration is believed to be essentially nonreversible — that is, a smoke particle which collides with the filter will not rebound and re-enter the smoke stream. A second type of filtration is the condensation of vapor from the gas phase of the smoke on the filter surface; this condensation is followed by adsorption or absorption of the component by the filter medium. Filtration by this mechanism is reversible; violatile components that condense on the filter surface can revaporize and re-enter the smoke stream.

Filtration can also occur by a process in which both the smoke component and the gas phase are involved. A smoke component may vaporize from the smoke particle and subsequently condense on the filter surface. This type of filtration is also reversible; the smoke component can vaporize from the filter surface and condense on a smoke particle and in this way re-enter the smoke stream. There is considerable interaction between the means of filtration, and there sometimes is no clear-cut division between various mechanisms.

Mechanical Filtration

The mechanical removal of a smoke particle is probably the most important mechanism in terms of reducing the total particulate matter in smoke; it is the means by which high-boiling smoke components are trapped. As shown in Figure 1, cigarette filters have an open construction, and in comparison with the size of smoke particles these openings are large. Filters do not act as sieves; sieve-like filters would have a tendency to clog and would offer a very high resistance to draw. Most of the filters used for filtering cigarette smoke particles have a fibrous structure, such as that shown in Figure 1. The fibers are not round, but there are longitudinal lobes and depressions along the length of each fiber. Peck[9] used a very ingenious method to demonstrate that the particles are trapped mainly on the outer surfaces of these lobes. He treated a filter containing smoke particles with methyl cyanoacrylate, which polymerizes on contact with water. A replica of the smoke particles is thus permanently fixed on the fiber surface and can be examined with an electron scanning microscope. Filtration of the particles depends on chance collision with the filter material and on the adhesive forces which hold the collided particles to the filter. Because of surface forces, the small smoke particles adhere to any surface they contact. Sinclair[10] and Rodobush[11] both estimated that 100% of the collisions of an aerosol particle result in retention of the particle. The efficiency of a filter in removing smoke particles depends, therefore, on the physical character of the filter material, and it is essentially independent of the chemical nature of the filter surface.

The mechanisms which influence the probability of a collision of an aerosol particle with a filter surface are: direct interception, inertial impaction, diffusion, physical adsorption, and chemical reaction. The effects of these mechanisms in the special case of cigarette smoke aerosols have been discussed by Dobrowsky[12], Fordyce[13], Keith[14], and others, and the following remarks are confined to some general conclusions that are fairly well established. Filtration of the larger smoke particles depends to a great extent on direct interception, whereas filtration of smaller particles depends principally on the mechanism of diffusion. Inertial impaction is believed to play a small role in the filtration of the very large particles. Gravitational forces are considered to be unimportant under the conditions for filtering cigarette smoke. There is some question about the role of electrostatic effects in filtration, but, generally, electrostatic effects are believed to be of little importance. The high moisture content of smoke is expected to dissipate electrostatic charges rapidly.

Since the efficiency of filtering the TPM is related to particle size, it is conceivable that selective removal of certain smoke components might be achieved by mechanical filtration if, as experimental evidence indicates, the chemical composition of the smoke is not homogeneous. Woodman[15], and his co-workers demonstrated that the concentrations of nicotine, scopoletin, and neophytadiene in a smoke droplet vary with the size of the droplet and that their concentration distributions are different. Mokhnachev[16] reported experimental evidence indicating the larger particles contain more water, carbonyl compounds, volatile acids, phenols, and paraffin hydrocarbons than the smaller particles; further, he showed that a filter trapped more of the larger particles than the smaller ones.

Woodman[15] clearly demonstrated that the particle-size selectivity obtainable with a conventional filter is not sufficient to effect substantial selective retention of smoke components. His work indicates that substantial selectivity cannot occur unless the different sizes of particles are present in widely varying concentrations, and filters can be designed which are extremely size-selective. Consequently, the feasibility of selective filtration by mechanical means is questionable.

Gas-Phase Filtration

The particulate phase — that is, the liquid and solid particles — makes up only 5 to 10% of the total weight of smoke. The remaining smoke consists of atmospheric gases, carbon dioxide, and other gases. Filtration of a smoke component which is present predominantly in the gas phase depends on mechanisms quite different from those that control mechanical filtration. The mean free path of molecules present in smoke is such that the probability of a gas molecule escaping a filter without making at least one collision with a filter surface is very small. However, a collision does not constitute filtration; rather, filtration of gas components is governed by the following mechanisms: adsorption (physical and chemical), absorption, and chemical reaction.

When a gas molecule strikes a filter surface, it may rebound into the gas stream, or it may be retained by either physical or chemical forces. Physical adsorption takes place through forces of physical attraction similar to those which cause liquification of gases. It is similar in nature and mechanism to the condensation of a vapor on the surface of its own liquid and is reasonably independent of the composition of the filter material. It takes place on any surface under the correct conditions of temperature and pressure. Most filters which remove the gas-phase components of smoke contain activated carbon with a very large surface area. With these filters, adsorption takes place by diffusion of the smoke component into the fine capillaries of the carbon. Chemisorption (chemical adsorption), on the other hand, requires electron transfer or electron sharing between adsorbent and adsorbate and, therefore, possesses a certain specificity which physical adsorption does not. For chemisorption to occur, the adsorbent's surface must be clean. The specificity obtained with chemisorption is due to the fact that not all surfaces, even when clean, are active in chemisorption. In practice, the requirement that the adsorbent be free of foreign material has so far prevented the successful development of cigarette filters based on chemisorption. A large amount of water in tobacco and tobacco smoke would be expected to deactivate chemisorption sites. Efficient filtration of a gas-phase component generally requires condensation followed by physical adsorption, absorption, or chemical reaction.

Absorption is an important mechanism by which certain smoke components are absorbed or dissolved by the filter medium. It occurs quite readily with cellulose acetate filters when the smoke component migrates into the fiber. It is a highly selective means of filtration, since it depends on the affinity of the smoke component for the filter surface. Highly selective filtration of gaseous smoke components can also be achieved with filters capable of reacting chemically with certain smoke components. Patent literature contains a considerable number of references to such filters.

Filtration of "Semivolatile" Components

A smoke component may be distributed between the gas phase and the particulate phase. Davis and George[17] called these components "vapors"; Waltz, Haüsermann, and Hirsbrunner[18] and Graham[19] labeled them semi-volatile smoke components. Although they do not constitute a large percentage of the total weight of the smoke, they are of considerable importance to smoke taste and aroma.

Three semivolatile smoke components with widely different properties are shown in Table II. Experimental data concerning filtration of these components serve to illustrate some aspects of the filtration process.

TABLE II

"SEMIVOLATILE" SMOKE COMPONENTS

Component	Vapor Pressure (50°C), Mm	Affinity for Smoke Particles
Phenol	2.3	High – constant [a]
Tetradecane	0.1	Low – decreases [a]
Nicotine	0.3	Medium – increases [a]

[a] Change in affinity as puff number increases.

In 1961, Hoffmann and Wynder[20] demonstrated that certain phenols were selectively retained by cellulose acetate filters by absorption of the phenols into the cellulose acetate fibers. Spears[21], Lipp[22], and others demonstrated that the degree of selectivity is a function of the filter material and could be altered with plasticizers or other liquid additives. Spears[21] suggested a mechanism by which the phenol is transferred from the smoke particle to the filter surface. More recently, Curran and Miller[23] demonstrated that although phenol, once in contact with a cellulose acetate filter, is not eluted or desorbed from the filter surface by passing warm air through the filter at standard puffing rates, it is eluted from the filter by a smoke aerosol. They also demonstrated that the amount of phenol present on the filter and the percentage which elutes are fairly constant as the cigarette is smoked.

Tetradecane, which has a lower vapor pressure than phenol, has a higher rate of elution from a filter than phenol, but this rate decreases as the puff number increases. The decrease in elution of this hydrocarbon is explained by the high moisture content of the aerosol in the latter puffs; the moisture reduces the affinity of the hydrocarbon for the smoke particles. Waltz, Haüsermann, and Hirsbrünner[18] showed that a substantial increase in the water content of smoke occurs as a cigarette is smoked.

Conversely, the rate of elution of nicotine increases with puff number, again probably because of the increased hydrophilic nature of the aerosol in the later puffs. Evaluations by Curran and Kiefer[24] demonstrated that 15 to 25% of the nicotine that comes in contact with a cellulose acetate filter is eluted by subsequent puffs.

SUMMARY

The filtration mechanisms operating within a cigarette filter involve direct interception, diffusion, inertial impaction, gravitational forces, and electro-static forces. Direct interception of smoke particles by fibers is an important mechanism by which smoke particles, particularly large particles, are removed. Inertial impaction plays a minor role but is responsible for removing many of the large particles. Diffusion is the major means by which the small particles are removed. Condensation on a filter surface is the first step in filtration of gaseous components. Condensation may be followed by physical adsorption on the surface or into the pores of adsorbents, such as activated carbon. Absorption may occur by diffusion into the molecular structure of the filter medium to form a solution.

Once a smoke component is captured by the filter, either by mechanical filtration or by condensation, it may be eluted during subsequent puffs. The probability of a component eluting or being desorbed from the filter is a function of the physical and chemical properties of the smoke component, the physical and chemical properties of the filter material, and the physical and chemical properties of the smoke aerosol.

REFERENCES

1. W. Humphries, Tob. Rep. , 97, No. 2, 21–26 (1970).

2. C. N. Davies, editor, "Aerosol Science," Academic Press, London and New York, 1966.

3. R. G. Dorman in "Aerosol Science," C. N. Davies, editor, Academic Press, London and New York, 1966, pp 195–221.

4. J. Pick in "Aerosol Science," C. N. Davies, editor, Academic Press, London and New York, 1966, pp 223–280.

5. I. Langmuir, Washington Office Tech. Service O. S. R. D. , Rep. No. 865, 1942.

6. W. Stöber, 5th International Tobacco Scientific Congress, Hamburg, 1970.

7. G. P. Touey and R. C. Mumpower, Tob. Sci. , 4, 55–61 (1960).

8. R. W. Jenkins, Jr. , R. H. Newman, R. D. Carpenter, T. S. Osdene, 5th International Tobacco Scientific Congress, Hamburg, 1970.

9. V. G. Peck, American Chemical Society–Canadian Society Symposium, Toronto, May 24–29, 1970.

10. D. Sinclair, "Handbook on Aerosols", U. S. Atomic Energy Comm. , U. S. Government Printing Office, Washington, D. C. (1950), pp 64–76.

11. W. H. Rodebush, Ibid. , p 118.

12. A. Dobrowsky, Tob. Sci. , 4, 126–129, (1960).

13. W. B. Fordyce, I. W. Hughes, M. G. Ivinson, Tob. Sci. , 5, 70–75, (1961).

14. C. H. Keith, 24th Tobacco Chemists Research Conference, Montreal, 1970.

15.   W. C. Owen, D. T. Wescott, G. R. Woodman, 23rd Tobacco
      Chemists Research Conference, Philadelphia, Pennsylvania, 1969.

16.   I. G. Mokhnachev, Summary of paper read by M. Haüsermann at the
      5th International Tobacco Scientific Conference, Hamburg, 1970.

17.   H. J. Davis and W. T. George, Beitr. Tabakforsch., $\underset{\sim}{3}$, 203-214,
      (1965).

18.   P. Waltz, M. Haüsermann, R. Hirsbrunner, 4th International Tobacco
      Scientific Congress, Athens, 1966.

19.   J. F. Graham, 22nd Tobacco Chemists Research Conference,
      Richmond, Virginia, 1968.

20.   D. Hoffmann and E. L. Wynder, Beitr. Tabakforsch., $\underset{\sim}{1}$, 101-106.

21.   A. W. Spears, Tob Sci., $\underset{\sim}{7}$, 76-80, (1965).

22.   G. Lipp, Beitr. Tabakforsch., $\underset{\sim}{3}$, 109-127, (1965).

23.   J. G. Curran and E. G. Miller, Beitr. Tabakforsch., $\underset{\sim}{5}$, 64-70, (1969).

24.   J. G. Curran and J. E. Kiefer, 5th International Tobacco Scientific
      Conference, Hamburg, 1970.